A Symphony
in the Brain

ALSO BY THE AUTHOR

Last Refuge:
The Environmental Showdown in the American West

The Open Focus Brain:
Harnessing the Power of Attention to Heal Mind and Body
(with Dr. Les Fehmi)

A Symphony in the Brain

∞

THE EVOLUTION OF THE NEW BRAIN WAVE BIOFEEDBACK

∞

JIM ROBBINS

GROVE PRESS
New York

Published simultaneously in Canada
Printed in the United States of America

REVISED EDITION

Library of Congress Cataloging-in-Publication Data
Robbins, Jim.
A symphony in the brain : the evolution of the new brain wave biofeedback /
by Jim Robbins.
p. cm.
Includes bibliographical references and index.
ISBN-10: 0-8021-4381-4
ISBN-13: 978-0-8021-4381-5
1. Biofeedback training. I. Title.
RC489.B53 R63 2000
616.8'046—dc21 99-086648

DESIGN BY LAURA HAMMOND HOUGH

Grove Press
841 Broadway
New York, NY 10003

08 09 10 11 10 9 8 7 6 5 4 3 2 1

For the children:

To my own, Matthew and Annika;

To Jake Flaherty and Brian Othmer, who helped show the way;

And to all who suffer needlessly,

this book is dedicated.

Contents

∞

Acknowledgments

∞

I knew little about the brain and even less about biofeedback when I started this book, so bringing me up to speed on those things required a great deal of patience on the part of those I questioned. I am especially indebted to Sue and Siegfried Othmer and Dennis Campbell, who indefatigably answered my questions and indulged my interest, long before a book was in the works. Barry Sterman, Joel Lubar, Michael Tansey, Margaret Ayers, Les Fehmi, and Susan Shor Fehmi also patiently fielded the dumb questions that journalists routinely ask. Gratitude is owed to Ray Flaherty, Lisa Larsen, and Jake Flaherty for sharing part of their lives with me. A thank-you is also due to the following: Bernadette Pedersen for hooking me to the brain wave equipment and buying lunch. Rob Kall was an intrepid guide through the labyrinth of personalities and protocols of brain wave training. Chris Carroll was extremely generous with his time and comments and suggestions, and helped me understand the intricacies of the scientific process. Anton Mueller was integral to the process of conceptualizing this book. My agent, Lisa Bankoff, took care of business. To my good friend Vaughn Sarkissian, CPA, for his excellent abilities to account. Morgan Entrekin for his commitment. Andrew Miller for judicious changes. Richard Krizan for his wise ways and manuscript review. D.D. Dowden for the right side of her brain. Thanks also to Jeanne Jiusto, Faith Conroy, David Spencer, and Florence Williams for their reads, and to Sara Luth and Randall Mann for a place to stay. Tony Jewett for the photo; and my parents, Jim and Betty. And as always to Chere and Matt and Annika for putting up with the long work days and travel that it takes to research and write a book.

May their brain waves often be synchronous alpha.

Preface to the Revised and Expanded Edition

∞

Ten years after I wrote my first article about neurofeedback for *Psychology Today* and eight years after the first edition of this book, neurofeedback has not exploded onto the treatment landscape, nor has the number of research projects grown exponentially.

Brain wave training remains a victim of the fact that it is outside mainstream concepts, is far ahead of the science of how it works, has a persistent but undeserved reputation as a softheaded "new age" idea, and is a model that—unlike the drug model—doesn't lend itself to astronomical profits.

The field, however, has gained a great deal of acceptance that it didn't have a decade ago. It has moved out of the small circle of dedicated practitioners who gave birth to it and nurtured it, refusing to let such a powerful technique disappear. Civilization is closing in on the lost tribe of brain wave trainers that I described in the first book. The argument that studies are lacking is becoming irrelevant: first, because there *are* more studies; second, because modern medicine is failing to deal effectively with emotional stress beyond the use of medication; and third, because a revolution is under way in which people are taking more responsibility for their own health. Many people are no longer passive consumers of health care but are thinking for themselves. They don't trust big medicine and big pharmacology to have their best interests at heart. Health care is being democratized. More people also believe that "absence of evidence is not evidence of absence": the fact that there are few rigorous studies of an idea doesn't mean the idea isn't powerful; it may mean that science refuses, for whatever prejudices, to study the concept. That is certainly the case with neurofeedback.

The evolution of the discipline continues. The technique is being used by a growing number of mainstream academics—for example, at UCLA, the University of Utah, and the University of Washington. Neurofeedback research and treatment have taken off in Germany, Norway, Switzerland, and other parts of Europe, where they are unencumbered by "flower child" connotations from the 1960s. The 2006 World Cup champion Italian soccer team, L. A. Clippers center Chris Kaman, and the Olympic gold medal skier Herman Maier are among athletes who have benefited from neurofeedback. There are even "brain hackers"—technophiles who have built their own systems to play with their brain waves. What neurofeedback can do seems even more promising than it did a decade ago, especially in the areas of autism, attention deficit disorder, attention deficit hyperactivity disorder, anxiety, chronic pain, and post-traumatic stress disorder. There is a new generation of faster, more modern, more powerful equipment and techniques. The technology is expanding beyond sensors on the scalp to functional imaging equipment and new sensors that read physiological signals at a distance. Siegfried Othmer, a founder of one of the first neurofeedback businesses, predicted ten years ago, "Someday neurofeedback will be adopted and no one will ever remember that they opposed it." The day is not quite here, but there is a palpable sense that the time is much closer at hand. I predict that it will not take another decade for the rest of the world to catch up. Neurofeedback is simply too powerful.

The field, a free-for-all of equipment, ideas, and approaches, is still a free-for-all but is at the same time hardening into an orthodoxy of its own, for good and bad. One leading part of the field is solidifying the science, making the approach uniform. At the same time, some critics say, it is making some of the same mistakes as modern medicine. It sees the client's brain as a machine to be tuned up to a normal range, rather than as a human being with complex biological and emotional systems and the capacity for transcendence and transformation.

Neurofeedback is also starting to incorporate other modalities. Many in neurofeedback saw the brain as the way into all of it. Of late, there is ample evidence to show that an integrated approach is vital, and signals from the heart, skin, and breath are being integrated. Nutrition, toxicity, and the emotional environment are a few other factors that some people say need to be part of the approach.

The field is still embryonic. Neurofeedback—or more specifically what it makes possible, the operant conditioning of autonomic function—still has vast untapped potential to help millions of people, many of whom are unaware that they are functioning suboptimally, who think they cannot be helped or who simply maintain themselves on a regimen of drugs.

Prejudices remain. Since this book first came out, Dr. Barry Sterman, whose pioneering research at UCLA on neurofeedback is unimpeachable, and who has published more than 150 papers in top journals, has applied for numerous grants to continue research. "But the National Institutes of Health will not give us grants," he said. "We've written solid grants but the minute you use the term neurofeedback certain people's minds snap shut. Sometimes I feel like Galileo."

* * *

This edition of *A Symphony in the Brain*, just like the first, is not meant as a manual to persuade people to visit a practitioner. It is a journalism study, undertaken without fear or favor, to swing a spotlight onto this remarkable phenomenon and show how much the field has accomplished, which is far more than enough to have science take a serious look at and enhance and explain the approach. There is no reason neurofeedback should not be taken seriously, save the rivalries and prejudices of science. "The literature suggests that EEG Biofeedback Therapy should play a major therapeutic role in many difficult areas," wrote Frank H. Duffy, MD, a Harvard-trained neurologist and associate editor of *Clinical Electroencephalography*, a peer reviewed journal not associated with the field. "In my opinion, if any medication had demonstrated such a wide spectrum of efficacy it would be universally accepted and widely used."

It's amusing and a little frustrating to watch neuroscience beaming over the discovery of "neuroplasticity," the fact that the brain and central nervous system are malleable—a discovery from the "decade of the brain," the 1990s—because in fact brain wave biofeedback practitioners discovered and harnessed this trait half a century ago.

It's never been a problem for me that neurofeedback doesn't, according to critics, have enough double-blind controlled randomized studies to show it works. There are plenty of treatments used by mainstream medicine that have no such studies. Instead, what concerns me now is what precisely neurofeedback is doing to the nervous system. It is powerful, often beyond belief; and though I think it is safe—far safer than most prescription medications—I would like to know more about what is taking place. For whom does it work best? For whom doesn't it work at all? What is the downside? And which neurofeedback is best for ADD, for example? Which is best for peak performance? Which is best for autism? Is there a material difference between healing by beta training and the healing that comes from the deep states of alpha?

Moreover, I would like to see the field integrate a more holistic approach to the human condition—body, mind, and environment. I've continued to write about neurofeedback for the last several years. Among other things, I researched and wrote a book with Dr. Les Fehmi on the original neurofeedback, alpha training called *The Open Focus Brain: Harnessing the Power of Attention to Heal Mind and Body*. As old as alpha training is, its effects remain fascinating and powerful and are still waiting to be discovered.

Open Focus not only offers life-changing experiences but also offers a new way of thinking about attention, awareness, and the place of human beings in the world. It is not only an alternative to prescription drugs and self-medication but an alternative to the way mainstream neurofeedback is going, the tune-up model.

Still, all neurofeedback takes us in the right direction. Since I wrote the first edition, even more doubt has been cast on the research behind antidepressants and stimulant medication, and suicide warnings are now required on these prescriptions. As biofeedback evolves, the "aha" moment will come when we as a culture realize we have a great deal of control over our nervous system and accept that responsibility. There is no reason for humankind to suffer from widespread anxiety, depression, ADD, ADHD, chronic pain, or a host of other ills. Most human beings—and this may be the most profound lesson of all from neurofeedback—are simply not inherently or irrevocably flawed. Instead, many—perhaps most—of the problems that plague humankind are a case of "operator error." We "own" our central nervous system to a far greater degree than we imagine. We can get our hands on the steering wheel and deal with anxiety, depression, ADD, and a range of other problems. Neurofeedback shows us how powerful we are.

Introduction

∞

> We have not been informed that our bodies tend to do what they are told *if we know how to tell them.*
>
> —Elmer Green

When I heard that there was a new kind of biofeedback that amplifies your brain waves and allows you to make your brain stronger, I thought, wasn't biofeedback something that came and went in the 1960s and 70s? I had never tried it, but I associated it vaguely with the seventies, the Beatles, and transcendental meditation. Biofeedback had a New Age whiff about it. Add the words "brain wave" and it sounded even wackier. Yet I was hearing interesting things about it, and I've always believed that the human mind is the last great frontier. I was battling chronic fatigue syndrome and had exhausted the traditional medical route, so I sold an editor on a magazine story about the new biofeedback and traveled to Santa Fe to test this "neurofeedback" and a variety of other technologies designed to enhance the performance of the brain at a weekend symposium put on by Michael Hutchinson, author of a book called *Megabrain.*

I hooked up to a neurofeedback instrument for my first session. After training for a half hour, my mind was tired, my thoughts muddled. But an hour or so after I finished, I experienced what is known as the clean windshield effect. The world looked sharp and crystalline, and I had a quiet, energetic feeling that lasted a couple of hours. It was the first time I had felt that way in years. And it convinced me to look a little deeper. This new biofeedback was something very different, I was told, a technique that could treat attention deficit disorder and closed-head injuries and depression and a long list of other problems. I looked into the research and found that the technique had been spawned by solid laboratory research on epilepsy in the 1970s and 1980s.

Still, the claims being made for neurofeedback seemed far too good to be true. If it was such a good thing, why hadn't I heard of it? Why hadn't it swept its way into the health care system? I've been blessed with a healthy streak of cynicism, however, and as a reporter, I know that the systems that surround us—science, health care, government, even journalism—function far less perfectly than is generally believed. Things fall through the cracks, get overlooked and ignored. It was no great leap to believe that something like neurofeedback could have been missed. And so I persisted, knowing from experience that these oversights are where some of the best stories dwell.

What I found is a small subculture of people who enthusiastically practice brain wave biofeedback: the simple science of quantifying subtle electrical information from a person's brain, amplifying it, and sharing it with that person, who can then control the information in a way that makes the brain more vigorous and able to do a better job of managing body and mind. Many of those who work in the field of biofeedback are passionate about what they do because they believe biofeedback is very effective, and will change the world. The more people I met in the field, the more impressed I was with their intelligence and commitment. Many had been using it for years—in some cases, two or three decades—and some of the results were astounding. At gatherings of neurofeedback practitioners, the stories of people who have had their memory restored, seen their child's hyperactivity or autism or epilepsy significantly improved, and had their lifelong migraine problem disappear are legion and routine. The effects of neurofeedback are not subtle. They are extremely robust. There is nothing else like it, not even other kinds of biofeedback. That's one of the reasons it has languished. There is nothing to compare it to.

Yet neurofeedback is neither miracle nor panacea. It is science. But because the science is young and relatively unknown, because it turns the way we have categorized and thought about illness upside down, because it functions outside of most frames of reference, it seems like mumbo jumbo. It works on a sound scientific principle, though one that was abandoned by the powers that fund science before it was fully investigated. A limited analogy can be drawn to acupuncture. There is no Western medical model to explain the technique, and it has long been dismissed in the West. But it works, and works well, and now Western medical science is grudgingly coming to terms with it and searching for a biological explanation. And many insurance companies pay for acupuncture therapy.

So I decided to tell the story of brain wave biofeedback. It is a journalist's dream. A sprawling, dramatic, multifaceted story, filled with controversial figures and tales of discovery, about a new technique that performs what most of us have been conditioned to think of as miracles. It has slumbered for more than thirty years, under everyone's nose. The most exciting thing is that it is only beginning to come into its own. "I feel like someone has given me a grand piano and I've learned to play a couple of keys," said Sue Othmer, one of the field's pioneers. I don't know if all of what practitioners claim will prove to be true. But there is enough evidence to know that neurofeedback has changed and will continue to change lives, a great many of them. It can treat serious problems that many people believe they must suffer with for the rest of their lives, without drugs or side effects.

The big question about neurofeedback is no longer whether it works. The questions are why it is as effective is it is, for whom, precisely, and how it can be made more powerful. There is something profound at work with neurofeedback. If the brain's multifaceted effort to create mind and run the body can be compared to conducting a symphony orchestra, its choice of music, its volume, and its tempo are all things we believe we are forced to accept, largely without question. That may no longer be true.

CHAPTER ONE

The Symphony

∞

For an eight-year-old named Jake the rest of the world has disappeared as he sits quietly in a darkened room and stares intently at a computer screen with a yellow Pac-Man gobbling dots as it moves across a bright blue background. A soft, steady beeping is the only sound. Jake is not using a joystick or keyboard to control the cartoon character; instead, a single thin wire with a dime-sized, gold-plated cup is fastened to his scalp with conducting paste. The sensor picks up the boy's brain waves—his electroencephalogram (literally, electric head picture), or EEG—and as he changes his brain waves by relaxing or breathing deeply or paying closer attention, he also controls the speed of the Pac-Man.

This is more than a game for the boy. Jake was born in crisis: he arrived more than three months before his due date, in July of 1990, and weighed just over a pound. He required open-heart surgery when he was three days old and spent the first two months of his life in an intensive care unit for infants. He survived, but with serious damage to his brain. The most severe symptoms showed up at the age of four when he entered his parents' room one evening drooling and unable to speak. He went into a grand mal seizure and fell unconscious on the floor. After that, the seizures came frequently, usually at night as he was falling asleep. Antiseizure medications blunted the severity of the seizures but could not prevent their onset. His parents, Ray and Lisa, kept an overnight bag packed for frequent trips to the emergency room, where the slight boy received injections of Valium to arrest the seizures. The sight of the needle going into their son filled them with apprehension. He also had small absence, or petit mal, seizures throughout the day, when his mind would go elsewhere, when he could neither hear nor speak for five or ten seconds. He was diagnosed with a speech problem and cerebral palsy, which diminished his fine- and gross-motor skills. Even at age seven, when I met him, he had not learned to tie his shoes, zip his zipper, or button his shirt. His learning disabilities were numerous and included attention deficit disorder and hyperactivity. He had speech problems and ground his teeth together constantly, something called bruxism. His sleep was troubled, and he often woke up ten or eleven times in the night. Despite this list of problems, there is a bright little boy inside of Jake, with a wonderful and sometimes peculiar sense of humor.

At the age of five, Jake started taking two heavy-duty antiseizure medications: Depakote and Tegretol. Both are depressants, both control seizures, and both have serious and worrisome side effects. The boy seemed logy and often tired. "We felt Jake was losing his personality," Lisa told me. "He was zoned out all the time."

I have known Jake's family since he was born; the incredible story of his birth made him something of a celebrity in our town of Helena, Montana. A local insurance company put his smiling baby picture up on billboards with the line "Baby Jake will always be special to Managed Care Montana," and talked about how its coverage had paid almost all of the approximately $350,000 in medical bills. On assignment in Santa Fe for a story about different technologies designed to enhance brain performance, I had heard about neurofeedback and the fact that its first and most effective use was with epilepsy. (Neurofeedback works on the same principle as other kinds of biofeedback except that it provides information about the brain, hence the prefix *neuro.*) At a Christmas party, I mentioned it to Jake's parents, who were eager to investigate an alternative to drugs. They researched the therapy on the Internet, made a series of appointments over a week, and drove three hundred miles to the nearest neurofeedback site in Jackson, Wyoming. They turned the week into a vacation, swimming in the motel pool, hiking in the Grand Tetons, watching elk at a wildlife refuge, and taking Jake to the local hospital for two one-hour "brain training" sessions per day on the computerized EEG biofeedback program.

Jake's brain has places where the electrical activity is not as stable as it should be. Research shows that the brain's electrical signals are subject to change and that people can be taught how to change them. All neurofeedback does is help guide the client to a specific frequency range and help him or her stay there. The brain does the rest. A technician has set the computer Jake is playing Pac-Man on so that when Jake spends time in those hard-to-reach frequencies, the Pac-Man gobbles dots and beeps like crazy. When he is not in those frequencies, the Pac-Man stops gobbling and turns black. Jake knows nothing about brain waves or his EEG, he simply knows that when the Pac-Man is gobbling and beeping, he is winning, and so he has learned how to adjust his brain waves to make the Pac-Man gobble dots all the time. It was easy: he caught on in just one session. As he spends more time in those frequencies his brain has trouble generating, his brain learns to function there on its own. This exercise makes the brain more stable.

It didn't take long for changes to begin to appear in Jake. "It took care of the teeth grinding within two sessions," Lisa told me when they returned from Jackson. "It took care of the sleep problems immediately." As the sessions continued, Jake became more settled, more centered. "We could carry on a conversation in the car on the way home for quite a while, the first time ever that we could carry on a two-way conversation for any length of time. His fine-motor skills improved, and he wanted to cut and draw and zip and button. He could never do any of that," Lisa continued. Unprompted, friends and relatives remarked that Jake seemed calmer and more centered. Later, Jake's parents repeated the protocol for another week. Again they noticed dramatic improvement. Jake went to see his pediatric neurologist, who had been skeptical at the outset, though he had signed off on the treatment. He examined the boy alone for twenty minutes. When he was done, he told Lisa and Ray that the treatment had indeed been effective. "Jake seemed more focused," Dr. Don Wight, the neurologist, told me later. "He could do things cognitively he couldn't do before the training. There was a qualitative and quantitative improvement in the way he was functioning. It was very real."

Jake's parents bought one of the $10,000 neurofeedback units from Neurocybernetics, a California biofeedback manufacturer, and have made it available to the community. Dr. Wight has been trained in the technique and has incorporated it into his practice. Jake has regular sessions with the local neurofeedback technician, Bernadette Pedersen, and continues to improve. In 1999, he received a three-year evaluation for his individualized education program in the public schools. "He had some phenomenal gains," said his mother. "He was an emergent reader going into second grade and after a year of steady training, he was reading at a fourth-grade level. One of the teachers called Jake's rate of improvement explosive, and I think it was."

Had Jake been born twenty years earlier, he would have had to live with his problems. But in the last decade this new treatment—called, variously, neurofeedback, neurotherapy, or EEG biofeedback—has dramatically changed the prognosis for Jake and thousands of other people. It is being used to treat not only epilepsy and learning disabilities, but also a long list of other problems that defy conventional treatment: cocaine, alcohol, and other addictions; vegetative states; serious and mild head injuries; autism; fetal alcohol syndrome;

discomfort from menopause and premenstrual syndrome; chronic pain; the symptoms of multiple sclerosis and Parkinson's disease; stroke; post-traumatic stress disorder; wild hyperactivity; Tourette's syndrome; depression; cerebral palsy; and much more.

All of this raises huge questions. What is neurofeedback? Where did it come from? What are brain waves? How can one tool treat so many disparate problems? How can something that works so well, and seems to perform miracles, not be in widespread use? Answers to those questions begin with an understanding of the three-pound organ known as the brain.

The history of efforts to unravel the source of human consciousness goes back thousands of years. Hundreds of ancient skulls with carefully drilled holes have been found in a variety of places around the world. Anthropologists have documented a belief by some native peoples that trepanation, or drilling a hole in the skull, combined with prayer and ritual, could relieve certain physical problems, perhaps epilepsy. At one archaeological site in France, one hundred and twenty skulls were found, forty of them with human-made apertures. Some people apparently survived the "operations," for new bone grew at the edges of some of the holes—which ranged from the size of a dime to nearly half the skull. In Peru, anthropologists examined well-preserved, three-thousand-year-old mummies found near Cuzco and found that 40 percent of them had trepanned skulls. Stanley Finger, a neurologist who has looked at the finds, has estimated that there was a 65 percent survival rate. Whether the holes were made in a ritual or a de facto "medical" operation is unknown, but the mummies provide the earliest known record of making a connection between a person's head and his or her behavior.

In Egypt, a painted papyrus illustrates that three thousand years ago Egyptians recognized that a blow to the head could impair one's vision or coordination. A blow to the left side of the head, according to the papyrus, affected the right side of the body, while a blow to the right side of the head affected the body's left side, a description that proved to be fact. It was the heart, however, that the Egyptians revered, as the dwelling place of the human soul. (For most of human history, in fact, a "cardiocentric" view has dominated.) After death, the Egyptians, practitioners of an elaborate funerary ritual, removed all of the organs from the deceased and stored them in specially made ritual jars, except for one: the brain was simply pulled through the nose and discarded. The Aztecs also believed the heart was the superior organ and that it governed feeling and emotion, though they believed the brain was important for remembering and for knowing.

Hippocrates, writing between 460 and 379 B.C. may have been the first persuasive proponent of the idea that the brain is the source of human intelligence. Building on the work of two of his teachers, Alcmaeon and Anaxagoras, he had the prescient idea that epilepsy was the result of a disturbance in the brain. He believed that the gray matter was the source of many other things as well:

> Men ought to know that from nothing else but the brain come joys, delights, laughter and sports and sorrows and griefs, despondency, and lamentations. And by this, in an especial manner, we acquire wisdom and knowledge, and see and hear and know what are foul and what are fair, what are bad and what are good, what are sweet, and what are unsavory. . . . And by the same organ we become mad and delirious, and fears and terrors assail us. . . . All these things we endure from the brain, when it is not healthy. . . . In these ways I am of the opinion that the brain exercises the greatest power in the man. This is the interpreter to us of those things which emanate from the air, when the brain happens to be in a sound state.

Hippocrates' view, however, was an anomaly, too far ahead of its time to be taken seriously. Aristotle, who came along several decades later, was a primary proponent of the heart-centered human, primarily because he had seen chickens running around after being decapitated. He had also touched both a human heart and a human brain shortly after the death of their owner. The heart was warm to the touch, while the brain was cool and moist, and so he reasoned that the brain was a kind of regulator that "cooled the passions and the spirit" and "the heat and seething" that originated in the heart. Aristotle was so well respected and influential that this view reigned unchallenged for centuries.

Galen, a physician to Roman gladiators and emperors in the second century, played a major role in the evolution of early thought about the brain. He believed there were four substances or "corporal humors": yellow bile, black bile, phlegm, and blood, which combined in a person's heart with "pneuma," a spiritlike substance. This solution traveled to the brain through a mesh of very thin tubes—which he called *rete mirabile,* or the miraculous network—and was then distributed to nerves throughout the body to produce behavior. Illness came from an imbalance in the fluids. Too much black bile, for example, led to depression and melancholy, while too much blood created a hot temper. The vital part of the brain, Galen claimed, were its ventricles: three hollow structures in the center of the organ that he believed contained this mystical animating substance. The fluid that created intelligence was found in the front ventricle, knowledge or mind in the middle ventricle, and memory in the rear chamber. (Ventricles do in fact exist; they are reservoirs for cerebrospinal fluid.) The rest of the brain, including the gray matter, was thought not to be critical. Adopted by the all-powerful Roman Catholic Church as the truth, Galen's "cell doctrine" reigned for fifteen hundred years, largely because, from the fourth through the fourteenth century, the church banned study of the human body. The dissection of human cadavers was penalized by torture or death, and the evolution of neuroscience virtually ground to a halt.

Then, in 1347, the Black Death seized Europe and killed a third of the population. The church's theories of medicine were proven woefully inadequate, and as a result the monopoly the church held on ideas about humans and their place in the world was broken. The Renaissance blossomed soon after, spurring a new burst of thinking about the human condition. By the sixteenth century, researchers were dissecting cadavers.

An anatomist named Vesalius may have been one of the first to question the cell doctrine. Because ventricles were similar in animals and humans, and animals were not capable of thought, he reasoned, how could ventricles be the source of thought? The difference between humans and animals, he believed, was a larger, more developed brain, and the true source of thought probably lay outside the ventricles. In the seventeenth century, Thomas Willis, an English physician, published a thorough text on the anatomy of the brain, in which he claimed that the brain itself, not the ventricles, controlled memory and volition. His work sparked a new way of thinking and would later convince researchers to abandon the cell doctrine.

Yet the cell doctrine survived for years after Willis's findings. René Descartes, the influential seventeenth-century French philosopher, is one of the most dominant early figures in the study of human behavior, and his influence still deeply impacts beliefs about the brain and the body, and even about reality as we know it. Descartes promoted the concept of dualism, the idea that mind and body are separate. He claimed that the ability to think was a gift from the Creator and the supreme aspect of human existence, while the body was separate and subservient to the mind, little more than a biological machine. His ideas were embraced by the church, and Descartes had laid the foundation for the next three hundred years of reductionism and the modern scientific method, which still dominates Western thinking. Nature is no more than the sum of its parts. Devoid of a soul, on death the human body and brain could be freely dissected and reduced to their component parts.

But the philosopher's work was not finished. If mind and body are separate, how do the two interact in humans? First, he said, involuntary movements were reflexive, an automatic response. Voluntary movements were a different matter. The spiritual belief of the day held that the body was animalistic, an unfit vessel for something so divinely elegant as the human spirit; so how did a godly spirit live in a body and run the show without becoming contaminated? And where? Descartes solved the conundrum of contamination neatly by claiming that the spirit entered the body by, and commanded the network of tubes and fluid from a single point: the tiny pineal gland, an organ in the front of the brain (named for its resemblance to a pine nut). Located there, the divine mind was almost completely untainted by the body; on death it simply floated out of the human "machine" and left it behind. Descartes chose the pineal because it occupied a central place in the brain, because it was near the senses, and because it was surrounded by cerebrospinal fluid, then still believed to be the liquid version of the animal spirits that allowed the body to move. Descartes's interpretation was the first attempt to assign a specific task to a specific part of the brain.

One of the first tools to come along to aid in reducing the universe to its component parts was the microscope. Chemical dyes, created for the textile industry, were used to dye slices of brain tissue for study under the newly invented instrument. It apparently didn't work well at first. Anton Van Leeuwenhoek, inventor of the microscope, looked at the sperm cells of dogs and cats and claimed that he saw microscopic dogs and cats, which he named "animalcules." It was a shared hallucination, apparently, for it was confirmed by other researchers. Improvements in the technology later dispelled that notion.

The microscope lent itself to the next evolutionary step in thinking about the brain, the school of localization. Researchers looking at cross sections of brain tissue noticed that different parts of the brain had different types and numbers of cells and asked whether the differences in structure, the "cytoarchitecture," pointed to a difference in function. Explorers of localization of function thought they did. Among the pioneers were Franz Josef Gall and Johann Spurzheim, who, in the late nineteenth century, hypothesized that every kind of behavior was represented in a specific region of the brain and that the organ was the source of the mind. They were right about that much, and far ahead of their time, but their work led them off into other, more fanciful realms. They hypothesized that a person's personality and mental traits depended on whether a particular part of the cortex was over- or underdeveloped. If one was lazy, the portion of the brain that governed "industriousness and responsibility" was weak; while the portion of the brain that governed mathematics was highly developed in people who were good with numbers. They went even further along this line and developed a "science" of phrenology—and as a result lost scientific credibility. Phrenologists claimed that differences in development of various regions of the brain caused bumps in the skull and that a personality assessment could be done by means of something called cranioscopy: feeling the topography of a person's head and comparing it to an interpretive chart of what each bump meant. It was the rage in the elite social circles of the time to have one's bumps read and one's character assessed.

Though Gall and Spurzheim were wrong about phrenology, they were right about functions being localized in the cortex. Their work ushered in the beginning of thought on how the physical attributes of the brain affect who we are.

Localization gained substantial scientific support in 1861, as a result of the research of a respected French physician named Paul Broca. Dr. Broca worked with a stroke patient who seemed to hear clearly but could answer any question asked with only a single word: "tan." After the patient died, Broca removed his brain and found a large lesion on the part of the organ called the posterior frontal cortex, on the left side of the head near the temple. Broca was fascinated, and a search turned up eight other patients who had similar language difficulties in the wake of a stroke, a handicap called aphasia. Seven were found to have similar lesions. Broca hypothesized that this small region of the left brain—now called Broca's area—enables humans to speak. His research rocked the medical world and kicked off a search for functions across the gray, convoluted landscape of the brain.

Not long after, a German neurologist named Carl Wernicke discovered another area of the brain involved in speech, farther to the rear of the brain than Broca's area. Wernicke also came up with a model of how speech is assembled by networks in the brain, a model that still holds up and provides some understanding into the complex nature of brain function. A sense of what a person wants to say arises in the form of an electrical pattern in Wernicke's area and then travels to Broca's area, where a vocalization program is formed. That program is then communicated to the motor cortex, which activates the mouth, lip, tongue, and larynx muscles to create speech.

In 1848, several years before Wernicke's discovery, a metal rod in the hands of a twenty-five-year-old Vermont railroad construction foreman named Phineas Gage set off some dynamite and added a new dimension to the concept of localization. Gage was tamping dynamite into a hole in some rock when a spark from metal striking stone ignited the explosive and turned the three-and-a-half-foot rod into a missile that tore through Gage's left front cheek and his frontal lobe. It kept moving and landed a hundred feet away. Despite the trauma, and profuse bleeding from the wound, Gage was sitting up in minutes after the accident and was fully conscious, though dazed. After the wound healed, he was physically fine; it was his personality that suffered. Instead of the polite, shrewd, and level-headed guy he'd been before the accident—described by his bosses as "the most efficient and capable man"—he'd become a foul-mouthed lout who swore all the time and couldn't hold a job. He eventually became a freak show attraction, showing off his wound and the tamping rod that created it. His experience led to the unheard-of notion that very tangible brain cells in the frontal cortex somehow govern something as intangible as the human personality. It afforded a glimpse at an answer to a seminal question: How much of the human mind is dependent on the tissue and blood in the brain? Apparently a great deal.

Around the same time there was a development that would play into a different kind of understanding of the human brain, one concerned with the electrical nature of human tissue. The first notion that human nerve impulses are somehow electrical in nature goes all the way back to 1791, when Luigi Galvani, an Italian researcher, published a paper on the subject. Using a hand-cranked generator to send a mild current through frogs' legs, he found that the current made the leg muscles contract, and he proposed an inherent electricity, an animating principle, that must exist in all living organisms. His work was not conclusive, but it began an important line of inquiry. The first conclusive evidence of electrical nerve impulses resulted from the measurement of a nerve impulse—also in a frog's leg—by a German physiologist named Emil Du Bois Reymond in the 1850s.

An English physician named Richard Caton was the first to discover that the brain generated electricity, using a device called a reflecting galvanometer to make the discovery in the 1870s and 1880s. The galvanometer consisted of a wire and coil that vibrated when small amounts of electricity were detected. The instrument Caton used had a small mirror attached to the coils, and a bright, oxyhydrogen lamp cast a narrow beam onto the mirror, which reflected the beam onto an eight-foot scale painted on a wall in a darkened theater. When a signal was stronger, the light shone higher on the wall. Caton touched the electrodes from his instrument to the exposed brains of rabbits and monkeys. When an animal moved or chewed food or had a light shone in its eye, there was a corresponding electrical spike. Thoughts, Caton noticed, also generated a charge. He hooked up a monkey and recorded the current associated with chewing. "If I showed the monkey a raisin but did not give it, a slight negative variation of the current occurred," he wrote. Almost offhandedly, Caton also detected the weak flow of current across an unopened skull: it was the first account of what would become the brain's electrical signature, the electrical encephalogram.

Before they learned how to measure the signals being emitted, researchers actually learned how to put electricity into the brain with great effect, and it became an indispensable technique for mapping brain function. Two German physicians, Gustav Fritsch and Eduard Hitzig, working at a military hospital, discovered they could electrically stimulate the brains of patients who had parts of their skull blown away in battle. Using a galvanic battery that delivered a tiny dose of current, Hitzig wired patches of exposed brain and found that stimulating the back of the brain, the occipital lobe, caused a patient's eyes to move involuntarily. The two men later experimented on live dogs to see which part of the brain corresponded to which voluntary motor controls. Stimulation was a vital tool that expanded understanding of the brain and quickly became more sophisticated. Two Englishmen, Charles Beevor and Victor Horsley, worked with an orangutan. They divided the brain into a grid of two-millimeter sections, and they numbered each square. Then they methodically stimulated parts of the animal's brain and created a detailed map of function—squares 95, 96, 121, and 127, for example, caused elevation of the animal's lip on the upper right side. Electrical stimulation of the brain created an intricate picture of the brain that remains an indispensable part of neuroscience to this day.

In the 1880s an Italian anatomist named Camillo Golgi developed a new stain that made nerve cells much easier to study under the microscope—a seminal development, for it had been impossible to enhance the microscopic cell without killing it. Using the new stain, a Spanish anatomist named Santiago Ramon y Cajal turned the nascent world of neuroscience on its ear: he discovered the brain cell, the neuron. Until then the human brain had been thought of as just a blob. He also described the basics of how the cells pass on impulses, by reaching out to the body or dendrite of an adjacent cell with a kind of cable called an axon. He went on to make several other major discoveries about brain cells, including the fact that nerve cells morph or change. As a person studies a subject or learns to play an instrument, nerve cells along a pathway involved in the skill make more connections with other cells. In 1906 he shared the Nobel Prize with Camillo Golgi.

Hans Berger was nineteen, and serving in the German Army, when the horse he was riding slipped down a muddy embankment and he was suddenly thrust into the path of an oncoming horse-drawn artillery unit. For a moment he thought surely his end had come. While the young soldier was shaken, he escaped injury. When he returned to his barracks, he found a telegram from his father, inquiring about his well-being, for his sister had had a disturbing premonition that he had been gravely injured. "This is a case of spontaneous telepathy in which at a time of mortal danger, and as I contemplated certain death, I transmitted my thoughts, while my sister, who was particularly close to me, acted as the receiver," the square-jawed, mustachioed Berger wrote near the end of his life in 1940. The experience was a defining moment for him. When he returned to college, he changed his major from astronomy to medicine. He eventually became a psychiatrist and continued to explore the possible physiological explanations for that premonition.

Berger's research led him to Caton's work, which he expanded on. At the end of the day, after his psychiatric work had been completed, the shy, fastidious man, who kept a schedule of his activities down to the minute, would retire to a laboratory and work secretly for a few hours with a primitive string galvanometer. He also experimented on patients who had had a piece of their skull removed for medical reasons, which made it much easier to access clear signals. Even though the brain is a constant electrical storm, its electrical potential is only about fifty millionths of a volt, a tenth of the voltage that is measured from the heart, and is difficult to measure. Like Caton, Berger made a beam of light vibrate with the electrical signal he detected, though instead of casting the beam on the wall, he directed it on a moving piece of photographic paper, which then created the wavy graph of the brain wave. In 1924 Berger recorded signals from an intact skull—the head of his fifteen-year-old son, Klaus. Using lead, zinc, platinum, and many other kinds of leads, Berger made seventy-three EEG recordings off of Klaus's head, the first published human electroencephalograms. Unsure of how precise his measurements were, however, he waited five years before he reported his results in a 1929 paper called "On the Electroencephalogram in Man." The first frequency he encountered was in the 10-hertz range, which at first was called the Berger rhythm.

Berger's discovery caused a minor stir in Germany, and the Carl Zeiss Foundation gave him a technical assistant and state-of-the-art equipment to replace the clunker he was using. Berger went to town, hooking up all kinds of people to see what electrical activity he might find going on inside their skulls. He hooked up his fourteen-year-old daughter and asked her to divide 196 by 7. The EEG clearly showed when the mental activity began and ended. Berger hooked up infants and found that there was no EEG until they were at least two months old. This was evidence, he said, that the brain at birth is incomplete. Someone suggested he hook up a dying man, but he rejected that as immoral. He did wire up a dying dog, however, and watched until the EEG flat-lined. He did EEGs on schizophrenics and psychotics and was disappointed to find that their EEGs were normal—their craziness did not show up in unusual electrical activity that he could measure.

Despite all of the research that has since been done on the brain, no one is sure exactly what functions the electrical activity represents. One scientist compared measuring the brain waves to listening to the sound of a factory through a wall. What is known is that the human brain normally operates within a range of from 1 to about 40 hertz. (A hertz is the number of cycles per second; the higher the hertz, the faster the brain wave.) The frequencies are clumped into categories that denote their characteristics. The 1-to-4-hertz range is called delta and occurs during sleep and some comas. Theta is the 4-to-8-hertz range. It's called the hypnogogic state and is a kind of consciousness twilight that occurs between being deeply relaxed and sleeping. Berger's 10-hertz discovery falls within what he called the alpha range, which is a relaxed but awake state, from 8 to 12 hertz. He also named the beta range, from 13 to around 30 hertz, which is the range of normal waking consciousness. The Greek letters are a holdover from the early days of EEG; many researchers say the Greek alphabet categories are too broad to be meaningful and that there is no reason to have the names; they argue for simply referring to states by their numbers. The designations persist, however.

Berger's ideas were largely ignored until 1934, when two well-known British physiologists, Edgar Adrian and B.H.C. Matthews, announced that they had replicated his measurements of electrical waves. Only six years later, Berger's life ended in tragedy. In 1938, as he made his psychiatric rounds, he received a phone call from a Nazi official ordering him to fire his Jewish staff members. He refused and was ordered to "retire." In 1940, deeply depressed, Berger committed suicide.

The 1930s were the halcyon days for another area of brain research, one that's been largely forgotten—electrical stimulation of the brain, or ESB. Passing small amounts of current directly into the brain shows the role that frequency plays in the operation of the body's "master control panel." Few modern books about the brain mention much about the role of frequency in the brain, because modern neuroscience is concerned almost entirely with the cellular level, with an emphasis on drugs to alter chemical flows in the brain.

In 1934, two Yale biologists, E. Leon Chaffee and Richard U. Light, published a paper on an experiment in which they implanted electrodes into different areas of the brain of several experimental monkeys that were kept in a cage surrounded by three large electrical coils. When the power was turned on and a current was sent through the coils, it activated the electrodes implanted in the brains of the monkeys and stimulated very specific parts of their brains. One of the monkeys had the wire implanted in the part of the motor cortex that governs the arm, and when the switch was thrown, the stimulation made the animal swing its arm wildly. In another monkey a different placement in the motor cortex caused "a series of chewing, tongue wagging motions."

Walter Rudolf Hess, a Swiss opthalmologist, was an important figure in the use of ESB to map the brain. He anesthetized cats and placed implants deep into their diencephalon, an area of the brain that regulates both the involuntary (autonomic) and voluntary nervous systems; then he studied their response to small doses of current there. At one particular site, "even a formerly good natured cat turns bad tempered," he wrote. "It starts to spit and, when approached, launches a well-aimed attack. As the pupils simultaneously dilate widely and the hair bristles, a picture develops such as is shown by the cat if a dog attacks it while it cannot escape." With a well-aimed dose of current at another site, he caused them to evacuate their bladders and put them to sleep. He also destroyed parts of their brain with a pinpoint of electricity to see which functions were affected. Then the cats were euthanized, the brains were thinly sliced, and, under a microscope, Hess studied precisely what parts of the brain had been destroyed. His work yielded valuable insight into the brain and built substantially on Wernicke's ideas—that many functions are not the product of a single part of the brain, but instead are governed by a network of sites within the brain, communicating in split-second electrical impulses. Hess also discovered that a tiny region deep in the center of the brain called the hypothalamus (part of the diencephalon) is an especially critical unit, governing essential regulatory functions such as body temperature and hunger and, in association with the pituitary gland, the endocrine system, which governs the production and circulation of the body's vital cocktail of chemicals.

Another pioneer in brain mapping, of particular interest to the evolution of neurofeedback, is Wilder Penfield, a neurosurgeon who was educated at Princeton, Oxford, and Johns Hopkins Medical School before going to Canada to be director of the Montreal Neurological Institute in 1928. Penfield's surgical specialty was the removal of brain lesions, or damaged tissue, that caused severe cases of epilepsy. To carry out such a procedure without destroying important parts of the brain, he needed a detailed map, and the only way to get that was to work with conscious patients. Penfield constructed a small tent that covered a patient's head as the patient lay on his or her side. Part of the head was anesthetized with Novocaine, and a tiny trapdoor was sawed into the bone of the skull, which Penfield swung open to expose the surface of the cortex, the top layer of the brain.

Among other things, Penfield mapped a two-inch-wide strip across the top of the skull, from the tip of one ear to the tip of the other, which he named the sensory and motor homunculus. *Homunculus* is Latin for "little man" and was so named because all the parts of the body that govern senses and movement are represented there—the human being in miniature. As a patient lay on the table, Penfield prodded the homunculus with tiny bits of current and carefully noted the response. When he inserted an electrode in the area that governs speech in one patient, for example, the man emitted a vowel cry as long as the electrode remained in place; when it was removed, the cry stopped. The electrode was placed in the brain again, in the same spot, and the vowel cry began again; the man was asked to stop making the sound, and he could not. Thousands of sites were mapped in the cortex in hundreds of patients. Penfield had an artist draw the homunculus as a little man, making the size of his body parts relative to the size of the area in the cortex that governs their use. Because humans are verbal, the little guy's lips, tongue, and pharynx are very large; his hands are also disproportionately large.

Enthralled by what he was discovering, Penfield continued to probe with the electrodes. Stimulating the part of the cortex that governed hearing, for example, elicited a variety of reports of sounds by patients: "a ringing sound like a doorbell," "a rushing sound like a bird flying," and "I hear a boom or something." Current in the part of the cortex where memories are stored caused one woman to hear familiar music, while another thought she saw people entering the operating room with snow on their clothes. Yet another very clearly heard her dead son's voice speaking loudly. All of these experiences were much more real than a mere memory, something akin to a very vivid dream.

In the 1950s, an ESB experimenter named James J. Olds stumbled onto the pleasure center in the brain of a laboratory rat while testing the animal with an electrode implant. Intrigued, he placed the rat in a Skinner box, a specially designed container for animal experiments with a lever in it. Once a rat learns that pressing the lever delivers a food pellet, the rat is operantly conditioned. But Olds didn't reward the animal with food; he wanted to see if the animal would press the lever for a shot of electricity to the brain. First, Olds placed the electrode in the sensory motor region and the rat responded by pressing the lever ten to twenty-five times per hour, which is not much more than the rodent might do randomly. As the wires were moved closer to the areas in the midbrain that govern pleasure, including sex, and areas concerned with digestion and excretion, the frequency of the lever pressing skyrocketed. By the time the wires were deep in the "pleasure zone," the animal was pressing the lever a whopping five thousand times an hour. Even food-deprived rats completely disregarded a bowl of their favorite chow in favor of a shot to the zone. Norwegian researchers Carl W. Sem Jacobsen and Arne Torkildsen at the Gauster Mental Hospital in Oslo replicated Olds's rat work in humans. They implanted an electrode in the pleasure areas of the brain and handed the button to the patients. Some subjects actually stimulated themselves into convulsions.

The most flamboyant of the ESB researchers, and a visionary of sorts, was a Spaniard by the name of José Delgado, a Ph.D. psychologist who conducted research on animals at Yale in the 1950s and 1960s. In one experiment Delgado placed an electrode in the amygdala—a small mass in the brain that governs fear—of a cat chosen for its friendliness. When the juice was on, the cat withdrew from humans and hissed and spat and threatened; when the current was turned off, the cat became friendly again. Monkeys stimulated in certain ways became much more aggressive and territorial. In one monkey, famous for biting its keepers, Delgado inserted a probe into the caudate nucleus. When the switch was on, the monkey was calm, and Delgado could insert his finger in the animal's mouth. When the switch was off, the animal returned to its nasty self and wouldn't let anyone near. The Spaniard's most famous experiment involved the use of what he called a "stimoceiver," a radio-controlled unit that delivered current to extremely small solid-state receivers implanted in an animal's brain. At one point, Delgado climbed into a bullring in Spain with his stimoceiver and a red cape. As the bull charged, Delgado flipped the switch, and the fierce animal, the implanted electrode activated, stopped dead in its tracks and quietly trotted away.

Delgado believed deeply in the power of well-placed electrodes, and in his book *Physical Control of the Mind: Toward a Psychocivilized Society*, he laid out a plan for a utopian civilization in which people would wear electrical devices to end depression, pain, anxiety, and aggression and to stimulate pleasure, the will, and the intellect. The times, he said, demanded such electrical evolution. "The contrast between the fast pace of technological evolution and our limited advances in the understanding and control of human behavior is creating a growing danger," he wrote. Delgado called his utopia an electroligarchy.

Delgado also foresaw a pacemaker that could be used to treat brain dysfunction—an invention that arrived not too far in the future. In the 1950s and 1960s, Robert G. Heath and Walter A. Mickle, working at Tulane University in New Orleans treating schizophrenics, were desperately looking for ways to alleviate their patient's illness. They implanted fifty-two patients with electrodes—not only in the cortex, as Penfield did, but also in the subcortical levels, the deep part of the brain where emotional regulation is centered. When that part of the brain was stimulated, patients reported a general feeling of well-being, and the symptoms of their illness decreased. Intractable pain disappeared, and people talked faster. Stimulation of the amygdala caused a feeling of rage or fear; stimulation of the hypothalamus produced feelings of anxiety and discomfort, and patients complained of a wildly pounding heart.

Heath found that a tiny electrode, inserted in the cerebellum in the back of the brain, and powered by a matchbook-sized battery implanted in the abdomen, could control serious mental disorders. The first patient to receive the implant was a young man who flew into uncontrollable, violent rages and had to be tied to his bed. A small hole was drilled in his skull, and the tiny electrode was slipped into his brain and fired to activate the pleasure center—located in the septal area and in part of the amygdala—and to inhibit the place where rage is centered—the other part of the amygdala, the hippocampus, the thalamus, and the tegmentum. The unit delivered five minutes of gently pulsed current every ten minutes and worked quite well. The man was untied, let out of his bed, and eventually allowed to go home. Things went peaceably for a while. Then one day the man went on a rampage in which he wounded a neighbor, attempted to murder his parents, and barely escaped being shot by the police. He was carted back to the institution and x-rayed. The wires from the battery to the pacemaker had frayed and broken. Heath reattached the wires, and the young man calmed down and went home again.

Pulse generators, very similar to Heath's, are state of the art now for the treatment of Parkinson's disease. A company called Medtronics, in Minneapolis, manufactures a neural stimulator that delivers current to the brain and greatly alleviates tremors caused by the degenerative neurological disease. A similar pacemaker built by a company called Cyberonics is used for treating epileptic seizures and for treatment of severe depression. In one study, fifteen of thirty severely depressed patients who did not respond to any medication reported they felt much better after wearing a pacemaker the size of a pocket watch that stimulated the vagus nerve in their neck, which in turn stimulated the emotional region of their brain. The stimulator is also being used to fight, diabetes and hypertension, alleviate the symptoms of stroke and closed head injuries, and enhance memory.

There is one other field of inquiry in neuroscience that figures directly in the field of neurofeedback. Unlike ESB, the concept of "neuroplasticity"—the idea that the brain is not static but capable, if given the right stimulation, of dramatic and long-lasting change—has only recently been widely accepted as scientific fact. It has fomented a revolution in thinking about the brain. For many years, the assumption that the structure of the brain, after childhood, was fixed throughout life was unquestioned. Injury or disease could diminish the brain's capabilities, but there was no way to enhance those capabilities. "Once development is completed," wrote Nobel laureate Santiago Ramon y Cajal, "the sources of growth and regeneration of axons are irrevocably lost. In the adult brain, nervous pathways are fixed and immutable; everything may die, nothing may be regenerated."

Ramon y Cajal, it turns out, was wrong. A growing number of studies show that the brain is capable of great change. Some of the more interesting research took place at a convent in Mankato, Minnesota, called the School Sisters of Notre Dame. Why, scientists wondered, were so many of the nuns free of such problems as senility and Alzheimer's disease well into their eighties and nineties, far beyond the proportion in the general population? Dr. David Snowden, a researcher at the Sanders Brown Center on Aging in Louisville, Kentucky, asked the nuns to donate their remains to science. It seems to him that those nuns in the order who used their head—taught or went to college or performed other intellectual work, including puzzles—suffered less senility or other degenerative brain disease than their counterparts who performed manual labor. After they died, he removed their brain and placed tissue samples from each under an electron microscope and noted dramatic differences between the two types of brains. Connections between the cells of the intellectually active nuns were more robust. One possible explanation is that exerting the brain in new ways through the course of life, he says, creates new neuronal pathways, more synaptic connections, and significantly more cortex—a bigger and better brain.

In the 1970s, a researcher in Canada who worked with white lab rats found that when he split a litter—brought half home and left half in their cage in the lab—those that stayed at his home learned much more adeptly than those left in the lab. Mark Rosenzweig and a graduate student, Bruno Will, designed an experiment at the University of California at Berkeley to test the notion that somehow spending time at the researcher's home stimulated the growth of brain cells. They divided a population of similar rats among three different environments. In a standard social environment, three or four rats were housed together in an average-sized and plainly appointed wire cage with food and water available. A lone rat was housed in an impoverished environment, a small, barren wire cage with dim light. A third population, twelve rats, was housed in the equivalent of a rat resort, a multilevel habitat with food and water and a number of objects to climb on, move, and hide under. Objects were continually added and removed. After three months in their respective habitats, the rats were tested on learning skills. Researchers found that those raised in the enriched environment were much better learners than those raised in the impoverished environment and could learn to navigate a maze much more quickly. When they removed the rats' brains, they discovered physiological differences as well, similar to those found in the nuns. The cortices of the enriched rats were thicker; the number of glial cells, which play a role in supporting neuronal activity, were greater; and individual neurons had more elaborate branches and an increased number of connections to other cells than those of the impoverished population.

Bruno Will followed up the experiments later at the Louis Pasteur University in Strasbourg, France, by placing rats with brain damage he created—he suctioned lesions on their visual cortex—into the three different environments. The results are testimony to the power of "environmental therapy": brain-damaged rats placed in the enriched environment recovered their eyesight much better than rats in the standard or impoverished environments. Will then took the experiments another step further. In a population of normal rats, he damaged the hippocampus, a portion of the brain that is critical to the formation and storage of memories. Despite massive damage to that region, exposure to swings, toys, and other enrichments for a month enhanced and sustained the recovery of most of the rats' memory. Rats with similar lesions placed in the other environments recovered their ability to remember only poorly or not at all.

A recent study, with a very solid design, seems to end for good the notion of a static brain. In November of 1998 the journal *Nature Medicine* published the results of a study conducted by a team of Swedes and Americans. Five terminally ill cancer patients at a Swedish hospital were injected with a fluorescent green chemical dye called bromodeoxyuridine. (BrdU, as it is known, is a chemical building block of human DNA. It is used in cells only when the cells are new and begin to divide, and not in existing cells. It is used in some cancer patients to monitor tumor growth.) Each time one of the five cancer patients died, one of the researchers, Peter Eriksson, rushed to the hospital. After a pathologist removed the brain and pulled out the hippocampus, which is deep within the temporal lobe, Eriksson injected it with a red dye marker that attaches only to neurons. When he looked at a slice of the hippocampus under a microscope, brain cells in the tissue of the hippocampus lit up in fluorescent red and green. The red dye that had just been injected meant the cells were indeed neurons; the green meant the cells had been created after the injection of BrdU, toward the end of the patient's life. Undifferentiated cells were continuing to divide, and to produce new, fully functional neurons as well, right up until the death of each of the five subjects.

The systems that govern the human brain are the most complex and compact on earth, and, even though more has been learned about the brain in the last thirty years than in all of human history, science has not come close to understanding how all the pieces fit together to create human consciousness. It's a riddle of our existence: "If the brain were so simple we could understand it, we would be too simple to do so," according to one observer. But a great deal is known. The brain weighs about three pounds, is about 90 percent salt water, and has the consistency of a ripe avocado. There are four distinct regions to the brain. The top layer is the cortex, and there, especially in the front of the cortex, is where reason, planning, writing and reading, and a host of other cognitive functions take place. The cortex is what makes us human, what distinguishes us from the rest of the animal kingdom, by mitigating our baser instincts. Unfolded, it would be about the size of a handkerchief, it ranges from one thirty-second to one quarter inch thick. It looks like the covering of a tree trunk, and the name *cortex* is Latin for bark. Beneath the cortex is the mammalian brain, or the limbic system, the portion of the brain that governs pain and pleasure, including sex, eating, fighting, and taking flight. Below that is the diencephalon, which regulates our sleep and appetite. The bottom layer is the primitive regulatory machinery, the reptilian brain. This region is concerned with function—breathing, blood pressure, movement, and body temperature.

The brain is a wonder of information processing. It would take a computer the size of some states to come close to matching the number of "bits" of information in the brain, yet you can hold it in one hand. Information comes into the brain from the external world through the senses and is converted to an extremely complex mix of electrical and chemical energy. The four parts of the brain, as well as myriad areas across the surface of the cortex, must "talk" to one another constantly, and the brain accomplishes that by means of its vast assembly of tiny electrical devices, the neurons or brain cells. Neurons are like microscopic batteries. Their membrane builds up a charge electrochemically and then releases it, over and over again, in the form of what is called an "action potential," a surge of voltage that propagates down the axon to where it terminates on other neurons. Cells fire in unison to create thought and movement, and information travels around the brain in networks. (An EEG electrode reads the activity of about one hundred thousand neurons.) How information—thought—is encoded in this electrochemical soup is a deep mystery. There are as many as one hundred billion of these long and spidery neurons, or cells, in the human brain, and each one may make from hundreds to hundreds of thousands of connections with other cells, which means a total of perhaps ten to a hundred trillion connections. These connections are at the heart of how well the brain functions. The development of the brain, from infancy to adulthood, is akin to ecological succession. The cells in an infant brain are like an open grassland. As the child learns, is stimulated and exposed to the world, the grassland begins to sprout thicker grasses, then small shrubs and trees that are separated from one another. Then the trees grow closer together and larger, and branches multiply. Finally, there is a rich, dense canopy of connected neurons, teeming with life—in the case of the brain, the life is electricity and chemicals that contain information. The richer the canopy, the more connections there are among the trees, the more fertile the habitat for the flow of information.

In very general terms, the brain works as follows. A person calls to you from across the street and says hello. The auditory areas of your brain fire to hear what the person says and to make sense of it. The visual cortex at the rear of the brain lights up so you can tell if you recognize the person. Memories start churning, so you can match what you see and hear against what you know. In order to form a response, thoughts are generated and speech areas begin to fire, which talk to the motor cortex and tell it to activate our lips and our pharynx and to wave our hand. The emotions of the lower level of the brain may activate if the person is a threat, owes you money, or is someone you love. Each of these functions is governed by an assembly of neurons, which generate charges that travel through clusters of cells at between one hundred and two hundred miles per hour, switching on and off with precision. The brain does all of this simultaneously, or in parallel, and in serial progression, as well. Picture the brain as filled with tiny lights, and everything a person does fires certain collections of these lights at different speeds and brightness. "We know that within the brain, a great many electric processes can be identified, each with its own limited domain, some apparently independent, others interacting with each other," said W. Grey Walter, an English scientist who was a pioneer in the electrical realm of the brain. "We are dealing essentially with a symphonic orchestral composition, but one in which the performers may move about a little, and may follow the conductor or indulge in improvisation—more like a jazz combination than a solemn philharmonic assembly."

This intricate symphony of consciousness is at play constantly when we are awake and engaged with the world around us.

In Jake's case, the symphony was largely intact but confused. The conductor was not doing his job, and the orchestra was playing too slow. Neurofeedback, the model holds, rouses the conductor and resets him to his appropriate speed. Once the conductor is back in form, the rest of the players fall into line. Whether the problem is autism, epilepsy, post-traumatic stress disorder, or any of a host of other maladies, the answer lies in resetting the conductor and appropriately engaging the orchestra with neurofeedback. The notion has yet to be accepted by the medical establishment, but its time may have arrived.

CHAPTER TWO

That Special Rhythm

∞

Tom was an eccentric young man, and it fell to Barry Sterman, who was a graduate student in psychology at the University of California, Los Angeles, to tutor him. Test scores showed that Tom, an eleventh-grader, was extremely intelligent, yet he could manage schoolwork at only an eighth-grade level. He had an odd, stiff gait, his face was pale, and he seldom showed emotion. When he spoke, it was in single syllables, such as "hi" or "yeah." A staff psychologist at the learning center had diagnosed Tom as having schizophrenic tendencies and warned Sterman that he could be delusional. This tutoring experience would ultimately lead Sterman to a series of discoveries that transformed our understanding of the power of the brain.

At the time, Sterman was in the midst of a biology class on endocrinology, the study of the body's pituitary, thyroid, and other glands and the essential functions, physical and emotional, that they fulfill. "I was studying for my midterm, and part of that was thyroid function, and all of this stuff was running around in my head," Sterman says. "So I came from class, took a look at this kid, and said, 'My God, he's a walking definition of hypothyroidism.'" He recommended to the young man's parents that Tom have a basal metabolic rate test. A few days into a prescription for thyroid medication, Tom walked in for his tutoring appointment and called out with a smile, "Hi, Barry! What are we going to be doing today?"

Sterman was floored, and as he recounts the episode that happened more than four decades ago, he is still in awe of the difference in the teenager. "He had color in his face, he was less rigid, and he talked to me like a regular kid. It was a defining moment. I knew then I had to study physiology if I was going to work with human beings." Psychology by itself, he felt, just wasn't enough, for the physiology of the brain and the body were somehow intertwined with thoughts and emotions. It was an idea that was ahead of its time, for the Cartesian notion of a separate mind and body was still firmly entrenched. After he received an undergraduate degree in psychology, Sterman studied neurology and psychology at UCLA for his Ph.D. and did his dissertation in sleep research. (Later, as part of his postdoctoral work, he taught psychology at Yale, in 1964.) After he earned his Ph.D., Sterman accepted a job as a sleep researcher at the Veterans Administration Hospital at Sepulveda, California. It was here, doing sleep research on cats and monkeys in a two-story, red-brick laboratory, that a couple of fortuitous accidents would lead Sterman to the discovery of the power of teaching the brain to produce certain frequencies.

Now in his seventies, Maurice "Barry" Sterman is an affable man with a keen sense of humor and a deep, raspy smoker's voice. His career is a distinguished one, and he's well respected by his peers. He is a professor emeritus in the departments of Neurobiology and Psychiatry at UCLA. He taught physiology at the School of Medicine and the School of Dentistry, and also taught seminars on inhibition in the central nervous system and a survey of the neurosciences at the university's prestigious Brain Research Institute. Sterman is also a member of the UCLA Academic Senate as a representative of the

Department of Anatomy. He has written or coauthored some two hundred scientific papers.

Sterman has a reputation in the field as a tough, no-nonsense researcher who does not tolerate claims or statements based on what he thinks is anything less than rigorous science. His work has stood out in a field that has been accused of poor research methods and making claims beyond what the science can support. And he is not shy about voicing his opinion; some say he's arrogant and something of an intellectual tyrant. He has stood up at scientific meetings, in the middle of research presentations, loudly interrupting and challenging the methods and results of colleagues.

Sterman's most valuable tool in his studies was the electrical "fingerprint" of the human brain, the electroencephalogram, or EEG. Decades of experience with the EEG have made Sterman an expert on the electrical signals of brain and mind. "I can tell from an EEG whether someone's paying attention, and if they are, if they are paying attention to me or what they did last night. You can tell whether someone is mildly retarded from an EEG. Or whether someone is hyperaroused and can't relax, because there's very little in the way of rhythmic activity in the EEG, very little alpha. Everything depends on the topographical distribution" of the electrical frequency.

In 1963 Sterman was using the EEG as he retraced the footsteps of Ivan Petrovich Pavlov, the Russian physiologist who, using dogs as subjects, discovered the principle of classical conditioning. The son of a priest, Pavlov did wide-ranging research, studying the central nervous system, the cardiovascular system, and especially the digestive system, which won him the Nobel Prize in 1904. Later, Pavlov was in need of saliva for his work on digestion, great quantities of it, and to produce it he stimulated the flow of saliva in a dog by injecting meat powder into the dog's mouth. Before long, Pavlov noticed that the dog didn't need the injection to salivate; the animal's mouth began watering in anticipation of the powder as soon it could hear footsteps in the hall.

Pavlov began studying reflexes. He divided them into two categories: Unconditioned reflexes are those that animals and humans come into the world with, which enable us to go about life. Sleeping and eating are good examples. Conditioned reflexes are learned responses. They are strengthened by use and diminished by disuse. The ebb and flow of these behaviors allow an organism to react to a changing world around it. Pavlov built a special laboratory in Leningrad to study conditioned reflexes in dogs, one that excluded all stimuli save those he introduced. In it he built a special two-part chamber. A dog was kept in a fixed, upright position by a harness, isolated on one side of the chamber, while an experimenter sat on the other side. Tubes led from the dog's glands into the experimenter's side of the chamber so he could witness secretions. Then the researcher introduced various kinds of sensory stimuli and measured the results. A tray with food on it was swung into the animal's view. A light was flashed into the dog's eyes. A bell was rung. Smells were released. A metronome was played at various speeds. Pavlov conducted hundreds of experiments. The one he is most famous for paired the ringing of a bell with the delivery of meat powder. After a while the bell would cause the dog to salivate even if the meat powder was withheld—a conditioned response. In another experiment he attached an electrode to a dog's hind leg and explored the animal's response to "negative stimuli" or pain. After a buzzer sounded, he would shock the dog. As with the bell and the meat powder, the dog associated the buzzer and the shock and displayed physiological signs of distress when the buzzer sounded alone. What, Pavlov wondered, would an animal do if he sent it mixed signals, sounding the positive stimulus, the tone that caused salivation, and then a short time later sounding the negative stimulus, which was associated with a shock?

At first there were two separate responses, salivation and then distress. As he moved the two contrary signals closer and closer together, however, the dog was confounded and responded in a totally unexpected way. As the stress became unbearable, the animal simply checked out of the experiment altogether and fell asleep. Pavlov called this response internal inhibition, which meant the animal could voluntarily close down its system to escape the stress.

A different kind of conditioning was described in 1911 by E. L. Thorndike, who locked a cat in a puzzle box. Once the cat figured out a way to open the box, it could undo the latch faster each time. Later, the Harvard psychologist B. F. Skinner built on the concept of conditioning with the creation of the Skinner box. A rat was placed in a box with a lever. The animal would press the lever a certain number of times by chance. If pressing the lever led to a food pellet's dropping out, the probability that the rat would do it again increased. If it led to a negative reward, the probability that the rat would do it again decreased. This added the element of active involvement and is called operant conditioning.

These are the origins of the behavioral school of psychology, the study of how organisms react to external events. It was an approach that disdained the subjective inner states of emotions and feelings and sought to bring measurement and objectivity to the study of the mind. Behaviorism is responsible for a great deal of our understanding of how people move about in the world. Conditioning is a fundamental element of human behavior. At a simple level, if a bakery makes our favorite kind of doughnut, says Sterman, we will go out of the way every morning to take that route on the way to work. If a cop gives us a ticket for speeding on the way to get the doughnut, it is hoped that the negative stimulus will condition us to slow down. Most theories of learning and parenting are based on behaviorism. "You take operant conditioning plus genetic endowment, you put them together, that pretty much explains who we are by the time we're twenty," Sterman says.

The application of behaviorism took off in this country as a result of World War II. A shortage of physicians led to the drafting of experimental psychologists to deal with the emotional problems of returning veterans. Once relegated to research labs, behavioral psychologists began to develop a practical application for their work.

Conditioning voluntary responses is one thing. The big surprise in behaviorism came in the late 1950s and early 1960s. Neal E. Miller, a researcher at Yale, proved that by using operant conditioning he could teach laboratory animals to alter their autonomic functions, functions that supposedly weren't changeable. His first experiment involved carefully observing a dog. Whenever the animal began to salivate on its own, he rewarded it with a drink of water. Eventually, the dog learned to salivate whenever it wanted a drink. Miller moved on to rats, to see if they could control their heart rate. To avoid the possibility that they were somehow using the muscles in their chest around the heart to slow the heartbeat, he injected them with a dose of curare, the poisonous plant extract used by natives in South America to paralyze their prey. Then he stuck an electrode into the pleasure center of a rat's brain, and every time it lowered or raised its heart rate—depending on what he was training the rat to do—he rewarded it with a dose of current. Within ninety minutes he had taught the animals to raise or lower their heart rates by 20 percent. Dr. Miller moved on to humans in 1969 and successfully taught a group of patients with chronic tachycardia, an abnormally fast heartbeat, to slow their hearts. Whenever a patient's heartbeat fell below a certain level, he was rewarded with a pleasant electronic tone. Further studies confirmed Miller's work, including one at Harvard Medical School, where researchers taught a group of male subjects to raise or lower their blood pressure. The reward was a five-second look at a centerfold from *Playboy* magazine. These discoveries would lead to the notion that people could control their own brain waves.

As a sleep researcher, Sterman was interested in the big question that plagued the field at the time: Was going to sleep something a person chose to do? Or was it an automatic shutoff, something the brain imposed regardless of choice? Pavlov's work indicated that it could indeed be a choice, for his confused dog had decided to nod off. This is what Sterman believed. To prove it Sterman set out to replicate Pavlov's experiments. He had one big advantage over the Russian: an EEG instrument, which provided detail on how the brain was responding. Was the EEG of nighttime sleep, for example, the same as "internal inhibition"? Or was it another kind of sleep altogether? This was pure behavioral research, an investigation into one of nature's curiosities with no particular goal or economic application.

In the fall of 1965 Sterman began the experiment, bringing thirty cats to his lab, where they were kept in cages and deprived of food. Cats, from the pound or laboratory supply houses, were routinely used in this kind of research for several reasons. One, they are euthanized by the thousand every year because there are so many. Two, they are inexpensive. But most important, it is because cats—as opposed to dogs—are all roughly the same size and have a uniform-sized brain, which makes an atlas of brain sites applicable for the whole species. It is much easier, therefore, to be sure the electrodes are implanted in the same place in the brain in all the cats.

The first part of the experiment was basic behavioral training. A cat was taken out of its residential cage and placed in a two-foot-by-two-foot, aluminum, soundproof experimental chamber. Every time the animal pressed a lever, a ladle would dip into a reservoir of milk mixed with chicken broth (cats don't find milk tasty enough to work for it alone), and the ladle arm would swing to a hole in the cage big enough for the cat to stick its head through and drink. "These were happy and healthy cats, and they loved working with us," Sterman says. "We'd let them out of their cage, and they would run down the hall to the training facility and say, 'Let's go to work.'" Among the researchers working with Sterman was Wanda Wyrwicka, a physiologist who had defected from Poland. Sterman was placing the cats in the cage and waiting for the animals to figure out that pressing the lever would deliver a reward. Nonsense, said Wyrwicka. That takes too long. "She said, 'Don't wait for the cat to figure it out,'" Sterman remembers. "She put the cat's paw on the lever and said, 'Here, stupid, press this.'" The cats learned quickly. The team fastened stainless-steel screws into the cats' skulls so they could attach leads to the EEG instrument, which read the brain wave activity across the cat's sensory motor cortex, the strip of the brain across the middle of the skull that governs the cat's motor activity and its sensory processing.

After the cats were thoroughly conditioned to press the lever and get their reward, a new element was introduced: a tone. If the cat pressed the lever while the tone sounded, the dose of chicken broth and milk would not be delivered. It had to wait until the tone stopped before it could press the lever and get the reward. So the cat would sit in the box, press the lever a couple of times, and get its reward. Then the tone would come on, and the cat would have to wait. Sterman expected the cat to go into a state of internal inhibition, possibly into a "microsleep," which should be reflected in the EEG. But the cat didn't go to sleep. The animal entered a unique state—it remained absolutely still, though extremely alert, waiting for the tone to end. It is the same state a house cat waits in, feigning heavy-lidded indifference, as a bird makes it way near enough to be pounced on. Accompanying this motor stillness was an EEG "spindle"—that is, the scribbled display of the brain's rhythmic electrical signal on the EEG paper that was unlike any seen previously. "It was a clear, rhythmic change," says Sterman. "It was fascinating. We had never encountered this EEG rhythm before, and it didn't exist in the literature of the time." Later, Sterman found passing reference to the frequency, but there had been no systematic study. Sterman named the frequency sensorimotor rhythm—a rhythmic signal peaking in the range of 12 to 15 hertz. It is a beta frequency, but over a specific part of the brain, the sensorimotor cortex, and so it is called SMR.

Sterman carried the experiment a step further, forging the first, crucial link between the neurological and behaviorist findings of the time. Could a cat be operantly conditioned to create this specific range of brain waves on its own, to willfully alter what was thought to be out of its control? Neal Miller's work indicated that it might be possible. Sid Ross, a technician who worked with Sterman, fabricated an electronic filter that isolated the 12 to 15 hertz in the EEG. The cat sat in the experimental chamber, but no light or lever was used in the conditioning experiment. If the cat created a half-second burst of the unusual 12-to-15-hertz frequency, its brain waves automatically triggered delivery of a shot of broth and milk into the ladle. Over the course of about a year, Sterman and his assistants trained ten cats an hour a day, three or four times a week, and the cats learned to produce 12 to 15 hertz at will. (The researchers also trained a group to inhibit SMR to see what effect a lack of the frequency might have on sleep.) To see the effect of such EEG training, Sterman studied the cats' EEG during sleep, because when a cat is sleeping there is no possibility that it is trying to please its handler. The sleep EEG had been profoundly altered. Sterman also noticed that the cats slept much more soundly with a marked decrease in the number of times they woke up.

The final step of an operant conditioning experiment is a process called extinction. If the animal is thoroughly conditioned and the reward is withdrawn suddenly and completely, the animal will perform the task repeatedly to regain the reward. Sterman stopped providing the broth and milk mixture. Sure enough, the EEG showed that the cats were producing more SMR than ever. "I like the analogy that goes as follows: Every day you stop at that favorite doughnut shop," Sterman says. "One day you're on your way to work and you get to the shop, and the door is locked. You pull on the door, knock on the door, and even go around to the back door and pull on it. You try very desperately to get into the place. That's what extinction is."

The results of the cat training were published in the prestigious journal *Brain Research* in 1967. Sterman repeated the study with eight rhesus monkeys, using Spanish peanuts instead of milk and broth, and those studies showed the same results.

While Sterman's research results were interesting, he had no idea if the SMR work had any application in the real world. He had answered a basic research question and created a mildly intriguing, if obscure, niche for himself as the first person to isolate the 12-to-15-

hertz frequency, observe its properties in cats and monkeys, and train the animals to produce it. No one, save a handful of other sleep researchers, would ever know about the concept of SMR.

Another coincidence, however, ensured that Sterman's research would in fact spread. In 1967, during the SMR research, Sterman got a call from a friend of his, a researcher named Dave Fairchild. It seemed that Gordon Allies, a drug researcher who had gone down in pharmaceutical history as the inventor of amphetamine, had obtained a contract with the army to research monomethylhydrazine, or rocket fuel. The United States was playing catch-up in the space race with the Soviet Union, but the rocket scientists had a problem. Monomethylhydrazine was highly toxic. When workers breathed in the fumes, or came in contact with the substance, it caused nausea, severe epileptic seizures, and eventually death. There was also concern that at lower doses it could bring about hallucinations or disruption of cognitive functions. Astronauts orbiting the earth during the Mercury program had claimed that they had seen natives waving at them as they flew over the South Pacific. That was, of course, impossible, for Earth was miles below. Was monomethylhydrazine leaking into the cockpit somehow? The Defense Department awarded a contract to Fairchild and Allies for research into the toxic effects of rocket fuel. Unfortunately, Allies had tested on himself a chemical compound he was developing, and it had killed him. Fairchild asked Sterman if he was interested in taking on the study, and Sterman accepted.

Sterman brought in fifty cats for the new study. The animals didn't run down the hall to work in this experiment. They were injected with rocket fuel, ten milligrams for every kilogram of body weight. Again, their brains were wired to an EEG to measure their reaction. A few minutes after the injection, all of the cats did the same thing: they vomited, made noises, salivated, and panted. Most of them went into grand mal epileptic seizures after one hour. Most of them, but not all. While a small part of the research population—ten cats—displayed all of the symptoms the other cats did, the onset of seizures was substantially delayed in seven of the population and never happened at all in the other three. "Boy, did we scratch our heads over that," says Sterman. "We couldn't figure out what the hell was going on. Answering that question defined the next ten years of my life. I forgot all about sleep research."

Through the course of science, a remarkable number of key discoveries have been abetted by coincidence. This was one of them. The seizure-resistant cats, it turned out, were animals left over from the previous study, the study in which the cats were taught to produce the sensorimotor rhythm. What Sterman had done by teaching the cats to produce SMR, he would come to realize, was to strengthen their brain function at the sensory motor strip, the same way a person builds muscle mass by repeatedly lifting weights. In medical parlance, their "seizure thresholds" had been increased; their brains were now functionally altered so as to resist the spread of slow theta waves across the motor cortex that caused seizures.

The study demonstrated a clear connection between mind and physiology. Simply guiding a cat into a specific mental state to strengthen a cluster of neurons in the brain in turn prevented motor seizures. And there was more than just a change in the brain; there were physiological changes from the top to the bottom of the cats and monkeys Sterman had studied. "Very specific, measurable changes," says Sterman. "In the brain, cell firing patterns changed and cells in the motor pathway reduced their rate of firing. Circuit patterns changed. And we found changes in the body. Respiration stablized. Heart rate went down. Muscle tone in antigravity muscles decreased. Reflexes diminished."

These animal studies clearly demonstrate, says Sterman, that the effects of SMR neurofeedback are physiological and not, as some critics say, a placebo. *Placebo* is Latin for "I shall please." It refers to a common phenomenon in the medical world in which people who are part of a test on a new drug, for example, unknown to them, are given a sugar pill instead of the drug, but their health still improves, sometimes dramatically. Somehow they have fooled themselves and gotten better on their own. Placebo effects can be dramatic but are transient. They usually occur at a rate of about 30 percent in a given study, but they can be higher. "But there is no placebo effect in cats or monkeys," Sterman says.

Sterman would have been content to keep experimenting with animals, he says, but Wyrwicka, his assistant, adamantly insisted that it was his moral responsibility to use the powerful technique on humans. Sterman wasn't sure that humans had the same signature rhythm. With the equipment of the time there was concern that the EEG wasn't advanced enough to accurately pick up brain waves through the human skull, and one couldn't put stainless-steel screws into the human brain. Sterman tried a different approach. A neurologist referred several patients to him who, because of cancer in the skull, had had a portion of the bone removed. Without interference, Sterman says, "Their EEG was *beautiful*," smiling at the clarity of the frequency as if he were recalling a breathtaking mountain vista. "And there it was—SMR." The existence of SMR in humans was confirmed.

Then another stroke of serendipity came into play. For the first human subject he would teach to produce SMR, Sterman had to look no farther than his own research lab: a computer coder who worked for a colleague. Twenty-three-year-old Mary Fairbanks, who had suffered from a major motor seizure disorder since she was eight years old, was the perfect test subject. Two or more times per month she went into severe grand mal seizures, violently shaking and passing out. Drugs could not control them. Over the years she had meticulously kept a log of her seizures, detailing their severity and frequency, and that documentation was invaluable. Her seizure disorder had also been thoroughly recorded by the National Institutes of Health and by medical researchers at the University of Wisconsin.

Epilepsy is an invasion of an unwanted frequency in the brain. When a person is walking, talking, and engaging in everyday life, the brain is operating in the higher frequency ranges, called beta, from

12 to 18 hertz. As stated earlier, in many ways the normal waking brain is like a symphony perfectly timed to create the simplest tasks. In epileptics a portion of the brain is unstable, or hyperexcited, and it can't resist as slow theta waves, in the 4 to 8 hertz range, start to creep in. This in turn recruits other areas of the brain to produce the abnormally low frequency. During an epileptic seizure it's as if all of the musicians in a symphony are playing at once, but without a score or a conductor. Normal motor function is disrupted. If the animal research held up, the brain would be made stronger with SMR, better able to resist the unwanted theta, and seizures should diminish or be prevented.

So in 1971 Sterman wired up his first human subject to a biofeedback instrument that he had instructed his technician, Sid Ross, to build. The unit was nothing more than a simple black electronic box with two lights on it, red and green. As in the device wired to the cats, a filter separated out the 12 to 15 hertz. When Fairbanks produced SMR, she was rewarded. Instead of a shot of chicken broth and milk in a ladle, however, a green light came on. When she was not in that range, and the low-frequency brain waves that caused the seizures were dominant, a red light came on. Her task was to keep the green light on and the red light off so as to encourage the high-frequency waves and simultaneously inhibit the low frequency. The SMR frequency was not really new to Fairbanks; like everyone else she passed through it all the time, spending a split second in it here and perhaps a few seconds in it there. What the biofeedback equipment did was help her dwell in that state for longer periods of time. And dwelling there is what teaches the cortex how to maintain stability.

For the purposes of measuring EEG signals, the human head has been divided into nineteen different sites in an international system of measurement called the "ten twenty" system. Sterman trained just two sites, called C-3 and T-3, which lie between the top of the left ear and the very center of the top of the head. Almost all neurofeedback done in the early days was carried out at C-3.

One hypothesis about what might be going on in the brain during neurofeedback has to do with the way the cells in the brain connect with one another. Since information travels along the branchlike connections between cells called dendrites, the denser and greater in number these connections are, the better the transfer of information. As frequency increases during a neurotherapy session and the brain is activated, more blood than usual streams to that area of the brain—the nutrients in the blood may be strengthening or reorganizing existing connections, which increases the cells' ability to self regulate. This is what many scientists think happens during any learning process. (Brain scans show that in people who go blind and learn Braille, the neurons in the area that governs their reading finger become more robust.) The neurofeedback model holds that the brain wave training increases the stability of that area of the brain as well as its flexibility, or its ability to move between mental states (from sleep to consciousness or arousal to relaxation, for example). It allows the players in the orchestra to play their parts better, to find the correct tempo, to come in on time, and to stop playing when they aren't needed. Since every aspect of a person is driven by an assembly of neurons, the healthier those neurons are, the healthier are the functions that they govern.

Describing precisely how it feels to be in the 12-to-15-hertz range in the sensorimotor cortex is difficult. "It's not relaxation, it's not just not moving," says Sterman. "It's when we will ourselves to be still. It's a standby state for the motor system. You might think of it as a VCR; it's a pause button." One man who did SMR training, a tennis player, said that producing the rhythm feels to him very much the same as that calm and vigilant split second when he throws a tennis ball into the air to serve and waits for it to fall back down far enough to hit. Imagining that situation, in fact, enabled him to produce SMR in the clinic. And the secret of neurofeedback is that, though it sounds complicated, producing SMR is a simple thing for most people to do. In fact, as with learning to ride a bike, it's easier to do than to describe.

In 1972 Fairbanks trained for an hour a day, twice a week, for three months on the prototype neurofeedback instrument, for a total of twenty-four sessions. "I was very skeptical," Sterman says. "But the results were remarkable. She went nearly seizure-free in three months. She ultimately got her driver's license. You can see a sudden decline in seizures due to the placebo effect, but if it is a placebo, you usually get breakthrough seizures again after some time. [That is, the seizures return.] For her to be seizure-free three months later was unprecedented in the history of her type of seizure disorder." Not only did the treatment work, it was amazingly robust. The researchers noted other changes in their subject, including a shift from being "a quiet and unobtrusive individual" to being more outgoing and "showing an increased personal confidence and an enhanced interest in her appearance." In hindsight, it was vital that someone who responded so well have been the first to try it. If it had been someone whose response was lackluster or nonexistent, Sterman might have ended his work on humans then and there. Sterman wrote a paper on the case, and it was immediately accepted by the top EEG journal, called *EEG and Clinical Neurophysiology,* and published the same year. Publishing in a major scientific journal is the last big step in a research project and includes a rigorous review by peers. Only those studies that have met strict scientific protocol are published, and acceptance is an imprimatur on the quality of the work. Sterman had finished the first leg of an important discovery.

The paper thrust Sterman into the limelight of behavioral studies. He had discovered a special rhythm: three notes that played a central role in brain self-regulation and could be trained to improve brain function. In 1973 several physicians and neurologists came to work with him; Dr. Joel Lubar, a professor of psychology from the University of Tennessee at Knoxville, who would later become instrumental in the field of neurofeedback, came to Sterman's laboratory at Sepulveda on a nine-month National Science Foundation fellowship to observe Sterman's work after he had replicated the epilepsy work in his own lab. Eventually, Sterman's results were replicated in several other laboratories. This successful exporting of the technique to other labs is critical in demonstrating efficacy.

Four epileptics were recruited for the next phase of research, and the results were very good, a 60–65 percent reduction in grand mal seizures. This time the results were published in *Epilepsia,* a top journal in the field of epileptic research, in 1974. The grant money rolled in, from the Neurological Disease and Stroke Branch of the National Institutes of Health, and the funding allowed Sterman in 1976 to expand the next test to eight subjects. He made the testing more rigorous. The three-year experiment was constructed with an A-B-A design. As in the first experiment, patients were trained for three months to increase their SMR waves and suppress theta or lower-frequency waves, which cause seizures. The number of seizures dropped dramatically, as predicted. After three months the protocol was reversed, though the subjects weren't told, and they were taught to raise the "bad stuff " and suppress the good. The subjects started experiencing an increasing number of seizures. Three months later the protocol was reversed again. "They reversed their seizures when they got their SMR up and their theta down," says Sterman. "We did all-night EEG recordings to keep consciousness out of the picture. And their EEG during sleep showed changes. And when we reversed, their seizures went back to baseline." All of these changes were made in secrecy—not even the people applying neurofeedback knew who was getting what—to keep any placebo effect out of the picture. An A-B-A design is the most powerful study design of all. The ethics that govern studies no longer allow A-B-A designs, however, because they do harm to a patient by first making them well and then taking them back into their illness. These results were published in *Epilepsia* in 1978.

Sterman was well along the way to a major discovery: a nondrug, nonsurgical method of treating epilepsy. He parlayed his latest results into another National Institutes of Health study, with twenty-four patients over three years. Again, the study was carefully and conservatively designed, with three groups of eight subjects: two control groups and one experimental group. If drugs cannot stave off severe seizures, the next step is often surgery, and Sterman's study group included people who were on a waiting list for anterior temporal lobectomy, a procedure that removes the seizure focus—a small piece of damaged tissue that is the source of the low frequency. One control group received no treatment; its members were simply given logs and instructions on how to record seizures. The other two groups were divided into pairs. One person in a pair—let's say Robert—was given true neurofeedback. He was part of the experimental group. His partner, Ralph, a control, was hooked up and thought he was getting true neurofeedback as well, but he was in reality responding to a recording of Robert's EEG signal, which researchers were feeding to him. It was what psychologists call yoked, or sham, treatment. No one, save the administrators, knew who was getting what. It is known as a double-yoked design, because both people are hooked, or yoked, to the same treatment. If controls such as Ralph, who were trained to someone else's EEG, reduced their seizures, researchers knew a placebo effect was at work. But the control group didn't. Those with the real feedback reduced their seizures, while those with the sham treatment did not. "It worked wonderfully," says Sterman. "We had several patients that went totally seizure-free. The results were unequivocal." A year later the average rate of severe seizures had dropped by more than 60 percent. And those waiting for surgery? "None of them ever went back," Sterman says. After the study was completed, those in the control groups were given true neurofeedback training.

I showed Sterman's papers to several psychologists, and all of them feel strongly that Sterman's work is first-class and say his experiments were well and carefully thought out. "It was an elegant design," said Dr. Chris Carroll, a psychologist in full-time private practice in Glen Cove, New York, and a specialist in special education."It was very well suited to the questions being asked, and is top-notch work." Carroll is very familiar with neurofeedback, for he uses it to treat clients. But he is also critical of the lack of double-blind, controlled studies in the field. Sterman's work is one of the few exceptions. If there is a criticism, it is that the sample size—a total of 37 people in Sterman's studies—was small. But Sterman's work was also replicated in several other independent studies in other laboratories. All together 174 subjects were trained, and 142 showed substantial clinical improvement. Five percent went completely seizure-free.

Dr. Alan Strohmayer, a Ph.D. psychologist with a private practice, an assistant research professor of neuroscience and neurology at NYU School of Medicine, and the former director of the biofeedback program at North Shore University Hospital in Manhasset, New York, also said Sterman's work was excellent. "There's no placebo in cats, no expectation, no wanting to perform for the doctor," said Strohmayer. "The cats statistically demonstrated control over their brain waves. It's the best evidence we have that shows we're capable of training someone else's brain waves."

As we saw with the electronic brain stimulation experiments in the last chapter, frequency has a dramatic impact on the brain, affecting everything from motor skills to feelings of pleasure and pain. The difference between ESB and Sterman's work is that in neurotherapy, people are learning to generate the electrical current on their own. A number of studies show that the thalamus is the generator of rhythmic electrical activity in the brain—the orchestra conductor. Sterman believes the epileptics he trained were learning, through operant conditioning, to control the thalamic generator in the same way that Neal Miller's subjects learned to alter their heart rates.

While Sterman did not deal directly with the patients very often, a psychologist named Robert Reynolds did. Reynolds is a Ph.D. psychologist in Connecticut, treating traumatic brain injury with neurofeedback. As a psychology student at Cal State Northridge in the 1970s, he worked for Sterman, giving patients a battery of psychological tests as they began the study and testing them again as they left.

"These people were having grand mal seizures, multiple grand mals a day," he said, and the family was usually grim when the person was entering the study. "I would see them four months later, and the family was completely turned around." Seizures had diminished, and "the people would be laughing and the family was happy. It was remarkable." Reynolds's experience with neurofeedback led him back to the field, and he now considers it the most important tool in his work with brain-injured and ADD patients.

The double-yoked study had proven the procedure's efficacy beyond any reasonable doubt, Sterman felt, and the next step was to explore long-term management of epilepsy with the technique, with an eye toward a clinical application, something that could be used in a doctor's office or hospital setting. In 1982 Sterman submitted another grant proposed to the NIH and was awarded $70,000 a year for three years. Toward the end of the first year of the study, a strange thing happened. Sterman received a letter from the NIH saying that the grant had been reviewed again and the committee wanted a double-blind aspect to the study. Such midstream changes were unheard of, but Sterman had no choice but to comply. Then, instead of signing off on the already approved grant, the committee sent it back for further review. Then Sterman received another letter, to the effect that the work had achieved its objective and no further research was needed. Sterman was floored. Funding had been pulled out from under him. In the middle of the project, the study was over.

Sterman still simmers about the episode and feels he was a victim of politics, a casualty of an assault by the medical community at NIH. "Doctors want Ph.D.s to be in the laboratory documenting procedures, documenting drugs, documenting the treatments they could apply. And here I was coming up with a protocol for a long-term treatment of epileptics. It was a turf battle. Pure politics." Almost all of behaviorism was abandoned in favor of pharmaceuticals in the 1970s, and biofeedback is barely a blip on the radar screen of modern medicine.

Conservative in his approach to research, Sterman still wanted more data on the technique. But without funding, there was nothing he could do. He boxed up his files on the SMR training in cats, monkeys, and humans and moved on to other projects, including research for the U.S. Air Force, studying the ways pilots pay attention to their flying, and helping design cockpits for maximum efficiency. Eventually, on a part-time basis, he treated some epileptic patients at Hollywood Presbyterian Hospital with the biofeedback technique.

Elizabeth Kim was among those with grand mal epilepsy treated by Sterman after he finished his research. She started her therapy in 1983 on a fee-for-service basis. Nearly 25 years later, she says that the neurofeedback training changed things for her. "It increased the quality of my life tremendously," says Kim, who is now 48 and in charge of donor recruitment at a Southern California sperm bank. After several dozen sessions at Hollywood Presbyterian Hospital, she told her neurologist she wanted to begin reducing her medication, and she did. At the same time, her seizures went from an average of one per month to four to six per year and were much less severe. "What was most noticeable," she says, "were the periods between the seizures. Within a month I may only have one seizure, but in that time I may get an aura and feel like I am going to have a seizure five or six times. It creates a lot of anxiety. I was working full-time, and I would wonder, 'Am I really going to have a seizure? Should I go home?' With biofeedback this was reduced tremendously. I didn't have the feelings that I was going to have a seizure and the anxiety that goes with it." Several years after she stopped doing neurofeedback, Kim was taking one medication instead of the three she was taking before therapy, and her seizures have dropped to one or two a year.

What Sterman and his research team apparently accomplished, even though it wasn't fully appreciated at the time—and still isn't — was enormous. If brain cells are indeed becoming stronger and growing new and permanent connections, as his model holds, he has gone a long way toward proving a concept that has only been accepted by neuroscientists in the last few years: that the brain is a dynamic and extremely plastic organ. He has also found a way to capitalize on that plasticity. He showed that by simply teaching a person how to nudge it in certain directions, the brain is capable of profound change. People with one of the most serious and disabling of medical problems—intractable epilepsy—could be taught to heal themselves. And it wasn't that difficult. Sterman had also provided tantalizing evidence—in the lab, under rigorous scientific conditions—that there was a way to harness the mind-body connection. Simply guiding the way someone thinks can change the structure of tissue in the brain and, subsequently, other key parts of human physiology. Neurons, the work demonstrated, are where mind meets body. The concept was revolutionary, a whole new paradigm. Yet except for a handful of people, the discovery dropped off the map.

Why? Part of the problem was that it had come out of psychology rather than the medical world, and the medical world wouldn't accept it. Another problem, according to Sterman, was the bad reputation he felt biofeedback had earned among scientists. While there was a great deal of careful biofeedback research being done by other bona fide research scientists, including Les Fehmi, Joe Kamiya, Barbara Brown, and others, people were making wildly speculative claims in the 1970s about alpha training with biofeedback, including claims that it provided a shortcut to feelings of transcendence and to enlightenment. Such things went over poorly in the scientific community, and every kind of biofeedback was tarred with the same brush. The bad reputation continues to dog responsible proponents of the practice to this day, but they have made strides toward overcoming it, and they believe fervently that acceptance is just around the corner. For it turns out that early researchers were on to something after all.

CHAPTER THREE

The Birth of Biofeedback

∞

In 1958 a graduate student named Richard Bach made history as the first person to control his brain waves, as part of an experiment designed by a psychologist named Joe Kamiya, who taught at the University of Chicago. Kamiya had gone to college during the heyday of behaviorism, yet he thought there was something wrong with viewing humans as nothing more than an accumulation of their genetics and their response to external stimuli. What about dreams and introspection? He remembered being deeply concerned as a young man with existential questions about the universe and his place in it. Behaviorists considered such thoughts irrelevant, nothing more than noise in the system.

The Japanese-American psychologist disagreed and wondered what verbal description people would give different states indicated by their brain waves. Could an EEG show when someone was introspective? The first order of business would be to condition people to produce certain frequencies so they could be studied. For the first experiment Kamiya chose alpha, in the 8-to-12-hertz range, which is the most prominent rhythm in the brain and easy to produce, and might, Kamiya felt, lie within the frequency range of introspection.

He designed a controlled experiment to find out if someone could distinguish between the different brain wave categories. It wouldn't be easy, for there is only a subtle difference between brain waves. Kamiya attached a sensing electrode to the left side of the back of Bach's head—the left occiput, which is where alpha is most evident. As the student lay in a darkened chamber, Kamiya watched his EEG and spoke to him by intercom. "Keep your eyes closed," Kamiya told him. "I am going to sound a series of tones." After each tone sounded, the subject was to guess whether he was in alpha. Kamiya knew from the EEG if it was alpha, but the subject did not. When he said "yes" or "no," Kamiya would answer, "Correct" or "Wrong," depending on whether he was in alpha or not. "He asked me how he should go about it," Kamiya says. "I said, 'I don't know,' but I asked him to pay attention to what state of mind he was in when the guesses were right and when the guesses were wrong."

On the first day of trials—which consisted of sixty tones and sixty guesses in a thirty-minute session—Bach got about half of his guesses correct, about the same that could be accomplished by flipping a coin. The second day he got 65 percent correct, "which still is not much to write home about," says Kamiya. "But on the third day he got eighty-five percent correct, and now it's getting kind of interesting." On the fourth day, the subject made a couple of mistakes at the outset and from then on got all the rest correct. Intrigued, Kamiya decided to keep going and kept sounding the tone. Before it was over, he had sounded the tone four hundred times and gotten a correct answer each time. It went on so long Bach began to think there was something fishy and deliberately gave a wrong answer. For the first time in several hundred trials Kamiya answered, "Wrong." To this day, Kamiya remains in awe of Bach's ability to identify alpha. "I like to say that the first subject I had was sent by God, because I never had another subject who became as accurate in guessing their brain state as this guy was. The result was so unambiguous that it encouraged me to keep going. If I had kept getting fifty percent and I had to run a thousand sessions to find this kind of success, I might not have continued."

In the second part of the study, the student was asked to enter into the alpha state when a bell was rung once and not go into the state when it was rung twice. Again, Bach was masterful. "He had perfect control," Kamiya says. The subject and some others were also adept in learning to vary their alpha, going from low alpha to high alpha and back as they wished. "They were able to enter and sustain either state upon our command," Kamiya says. Others could not control it at all. It was the first controlled experiment to demonstrate that brain waves, normally thought of as involuntary, were subject to voluntary control. It launched the field of brain wave biofeedback.

Kamiya asked his prize pupil how he did it, but the young man said he wasn't sure. After more trials, however, he told the researcher that he found himself in the alpha state when visual images were absent from his mind. "When I imagine music from an orchestra, I'm in alpha," he said, "but when I imagine a visual image of the orchestra, it seems to cause an absence of alpha." Others described alpha as "letting the mind wander," "feeling the heart beat," and "not thinking."

Kamiya didn't see the experiments as significant at the time. "We were just curious to see if people could control their brain rhythms," he says. "We had no intention to help the ailments of mankind." Kamiya's work did not become well known until an article about it was published in *Psychology Today* in 1968, after he had moved to the Langley Porter Neuropsychiatric Institute in San Francisco.

As a general concept, biofeedback has been around a long time. It's a way of gaining information about the body to better manage it. Looking at a reflection in a mirror to comb your hair or put on lipstick is getting a very basic kind of feedback. Using a thermometer to take your temperature and to help you decide whether to take an aspirin is a kind of biofeedback. The first known record of biofeedback, or "self-regulation," with instrumentation to control part of the body not normally under voluntary control was in 1901. A man named J. H. Bair wrote a report called "Development of Voluntary Control" in which he described a mechanical device he had invented that allowed people to gain control of the muscles that wiggle the ears. He designed it, he said, "because of the light its solution would throw upon the nature of the will." The first brain wave biofeedback was reported in 1934 by E. D. Adrian at Cambridge, who with his partner, B.H.C. Matthews, had replicated Hans Berger's work, as mentioned in chapter 2. Sitting in front of an oscillograph and an amplifier that played a beat that reflected the frequency of his EEG, Adrian found that he could create the alpha rhythm at will "as soon as the eyes are closed, and maintain it with rare and brief intermissions as long as they remained closed."

The field of modern biofeedback that was born in the 1950s and 1960s was the outgrowth of two developments. First was the burst of evolution of electrical instrumentation that grew out of the war effort during World War II. Equipment up until that time wasn't sensitive enough to measure the body's faint electrical impulses very accurately. Second was research on stress and the role it played in illness. Walter B. Cannon was one of the first physiologists to study the powerful and long-lasting effects of stress on the human body in the early part of the twentieth century. (*Stress* comes from the Latin word *strictus,* which means tight and narrow.) In his lab at the Harvard Medical School, Cannon fed a cat food that was laced with a radioactive element called barium, so he could observe the cat's stomach with an x-ray. As long as the cat was content, its stomach muscles digested the meal in wave-like motions. When Cannon provoked the animal to anger or frustration, however, the stomach immediately halted its digestive motions and froze, even if the stimulus was removed, for an hour or more. The cat's body stopped nonessential activity to prepare to defend itself or to escape; Cannon dubbed the response "fight or flight."

Cannon also studied the profound effects the hormone adrenaline has on the human body. Just a couple of drops dissolved in

100,000 parts of water and injected into the cat caused it to arch its back, bare its claws, and dilate its eyes. Heart rate and breath rate, the amount of sugar in the blood, and blood pressure all increased. Cannon collected urine from members of the Harvard football team during an exciting game and found the same thing: a detectable increase in adrenaline and in sugar. Subsequent studies have shown that even being asked to solve a complicated math problem can evoke a milder version of the fight-or-flight response.

Another pioneer in stress research was Hans Selye, a researcher at McGill University in Montreal. In the 1950s Selye wrote a best-selling book called *The Stress of Life* and introduced what he called general adaptation syndrome (GAS). There are two components to the autonomic nervous system—the parasympathetic, which dominates during relaxation, and the sympathetic, which is switched on in reaction to external stress. Humans, Selye said, suffer from chronic activation of the sympathetic nervous system.

The first part of GAS is called the alarm stage. In response to a perceived threat—a car swerving toward you, a stranger walking quickly up behind you on a dark, lonely street—your body prepares itself to fight or take flight. Stress chemicals flood the body, making the heart beat faster, constricting blood vessels in the limbs to keep blood in the trunk, increasing breathing and perspiration, tensing muscles. When the situation calms, you go on about your business. The problem is that the body and mind often do not completely return to their original state; they've been permanently changed by the event. Capillaries throughout the body have constricted, heart muscles stressed, and muscles have retained some tension. This is viewed as remnant behavior, a holdover from the days when humans in a wild environment needed a surge of energy to deal with an attack by an animal or an enemy tribe.

The second part of GAS is called adaptation. After a while, a person gets used to the stress, even though the body and mind have not returned to their original state. If coping is not successful, the stress continues to exact a toll, to cause problems in the body, even though the person is unaware of it. That sets up Selye's final stage: exhaustion. The list of symptoms that may be related to stress-induced exhaustion is a long, seemingly all-encompassing one: anxiety, a variety of pains, depression, headaches, stomach and bowel problems, asthma, coldness in the extremities. Heart disease has long been associated with stress, especially in about 20 percent of the population known as "hot reactors." During the stress of daily events, the blood pressure of such people can shoot from a normal resting level of 120 to a whopping 300 and can eventually lead to a heart attack or stroke.

Stress can also dramatically impact immune system function. Robert Ader, a researcher at the University of Rochester School of Medicine and Dentistry, was one of the first to demonstrate that when he conducted a study of conditioned taste aversion on lab animals. The question was, Could a lab animal be taught not to like a favorite food? The answer was yes. Ader found that if he fed an animal sugar water, which it loved, and then fed it a nausea-inducing substance, the animal would no longer drink sugar water. But the really interesting discovery was accidental. By chance, Ader used an immunosuppressant—a drug that is used in transplant patients to keep their immune system from attacking their new organ—to cause the nausea. He later found that animals that had been conditioned to pair the taste of sugar water with immunosuppression had a much higher mortality rate when they drank sweetened water alone. The animals had been conditioned to diminish the function of their immune systems, which demonstrated a direct psychological effect on immunity.

What's been added to the body of knowledge about stress in recent years is the profound impact that emotions have on the nervous system. "The prime directive of the brain," writes Dr. Bruce Perry, a trauma specialist, "is to promote survival and procreation. The brain is 'over determined' to sense, process, store, perceive, and mobilize in response to threatening information from the external and internal environments. All areas of the brain and body are recruited and orchestrated for optimal survival tasks during the threat. This total neurobiological participation is important in understanding how a traumatic experience can impact and alter functioning in such a pervasive fashion. Cognitive, emotional, social, behavioral and physiological residue of a trauma may impact an individual for years—even a lifetime."

A stress chemical called cortisol is emerging as a primary player in damage to the brain. "It is the master stress hormone," said Dr. Ned Kalin, a psychiatrist at the HealthEmotions Research Institute at the University of Wisconsin, where they are studying the effects of stress on the brain with state-of-the-art brain imaging and molecular techniques. "In low doses it alerts us and organizes our behavior so we make sure to protect ourselves." But in higher doses, he said, "it leaves us stressed out, inattentive, disorganized and depressed. Severe stress affects the size of the structure [in the brain], cell death, and the number of connections between brain cells. And earlier in life the brain is much more vulnerable to insult." Topping the list of stressors, as numerous studies attest, is an absence of solid, caring relationships. "A car wreck is bad," in terms of stress, said Kalin, "But its not as bad as being neglected, isolated, or ostracized by your peers. Deprivation—lack of love, comfort security—is very stressful and can have big time effects." A study at the University of Minnesota showed that children who have a poor emotional attachment to their parents get higher rushes of cortisol during even mildly painful events, such as being vaccinated, than do children with strong parental bonds.

Research has shown that sustained exposure to cortisol can cause serious damage to the hippocampus, which affects memory, mood regulation and interpretation of space. Some researchers believe cortisol may also cause damage to other parts of the brain, notably the left prefrontal cortex. This region of the brain, right behind the forehead, is vital to humans, orchestrating emotion, arousal and attention and providing a restraint mechanism that keeps people from acting on impulse. It is key in teaching a child to feel remorse and establish a conscience. In fact, the prefrontal cortex is known as the

"Organ of Civilization" because it is largely what distinguishes us from animals.

According to work at the University of Wisconsin and elsewhere the left prefrontal cortex plays a key role in integrating positive emotion into people's lives. The tiny bit of tissue drives the networks in the brain that makes us feel good, while the right side drives anger, fear and other negative emotions. When the frequency of the two sides isn't balanced and the right side is higher than the left, people can't engage their positive emotions and become depressed.

Some research has shown stress has an extremely deleterious effect on the prefrontal cortex. Bruce Perry, a researcher at the Baylor College of Medicine has done brain scans of children who have been severely neglected. Key portions of their brain, he says, don't develop properly and are nearly a third smaller than the brain of a normal child of similar age. And with key areas of the brain damaged, problems arise. Adrian Raine, a psychologist at the University of Southern California, did a study in 1999 of twenty-one men with psychopathic personalities who had committed violent crimes. All of them had 11 to 14 percent fewer nerve cells in the prefrontal cortex—about two teaspoon's worth.

If key parts of the brain, or the entire brain, are weakened by stress chemicals, it could have major effects on the body as well. A common denominator in everything from chronic pain to immune dysfunction to heart disease and depression may be in the brain. "Many of these systems are regulated by the brain," said Kalin. "Hormones are regulated through the central nervous system for example and the brain, through the peripheral nervous system regulates pain responses, heart rate, a whole bunch of things." It seems that when stress damages the brain, it is analogous to damaged lines of code in computer software—neither body nor mind will operate correctly until the damage is fixed.

Where does stress come from? Some of the first exposures are the fearful or anxious experiences children have. An infant that cries unanswered, for example. Children who are yelled at or nearly hit by a car or forget to bring their homework to school or are physically abused or feel unloved all suffer varying degrees of psychological and physiological stress. It doesn't matter whether the threat is real or imaginary; the physiological response is the same.

Stress continues throughout life. Researchers Raymond Cochrane and Alex Robertson have compiled a Life Events Inventory and numerically ranked the stressful events that people experience, with such things as unemployment and a prison sentence at the top of the list, while going on vacation and getting new neighbors are near the bottom. There are also background stressors, those that are chronic and repetitive: traffic, noise, hectic schedules, family and job problems.

Biofeedback researchers have, over the years, developed a set of tools that measure stresses in different parts of the body and teach the client to reverse them by training various parts of the body to relax, many of which were thought for a long time to be beyond conscious control. In the 1960s in a book called *Muscles Alive,* John Basmajian, a researcher at Harvard, described an experiment he had conducted. He was studying cells in the brain's motor cortex, which send the message to muscle cells in the body to fire. An organized nerve path in the motor cortex is called a motor unit, and a single motor unit that travels down the spine might control anywhere from a few muscle cells to hundreds of them. Basmajian chose a motor unit that controlled a few cells at the base of the thumb. He inserted a tiny needle electrode into the thumb muscle. No one could see any movement of the muscle, but half of Basmajian's sixteen subjects were able to control the firing of a single motor unit in the brain that governed that muscle. The electrical impulse sent by the motor unit to the thumb was picked up and amplified over a loudspeaker. Each time a subject fired the small cluster of nerve cells, a *click* sounded over the speaker. The subjects learned to play distinctive tattoos by firing the brain cells once, twice, three times. They could imitate the sound of galloping horses and drum rolls on command. It was a finding far ahead of its time—the subjects had easily learned to handily control a tiny group of cells by use of their will. Basmajian asked them to describe how they did it. They couldn't, they said—they just did it.

Basmajian and a few other scientists went on to create the primary tool for traditional biofeedback, the electromyograph (EMG). People tend to hold chronic tension in certain muscles. Since tense muscles have a higher electrical reading than relaxed ones, small sensing electrodes are attached to tense muscles. The frontalis, the forehead muscle, for example, is a primary target for EMG, because when people concentrate hard or worry or suffer emotional distress, this is one of the muscles they often tense. The equipment displays the measurement on a computer screen and tells clients which of their efforts to relax the muscle are successful. EMG is used primarily for headaches and stress. Pelvic muscles are also trained with EMG to combat incontinence.

Hand warming is another long-time biofeedback task. A thermometer is attached to a finger; when the client successfully relaxes, the temperature on a digital readout rises or a tone beeps more slowly. Below-normal temperatures usually occur in people with anxiety, migraines, menstrual problems, and stomach disorders, and once a person learns to warm his or her hands, many of the other problems are alleviated or lessened. Another common type of feedback is galvanic skin response, or GSR. Stress causes increased perspiration, which is salty and a conductor of electricity. A fingertip sensor is placed on the hand, and as the client lowers his or her stress level, a beeping sound decreases. There are several other kinds of biofeedback, which measure heart rate, respiration, and other indicators.

Despite solid scientific credentials, biofeedback is an ungainly stepchild in the medical world, primarily because the biofeedback model is very different from that of allopathic medicine. When health problems arise, most people are accustomed to having something done to them: they are given a pill, subjected to an x-ray or an MRI, have their gall bladder taken out, or have new arteries made for their heart. Biofeedback requires the patient to take some responsibility for his or her own health; it is about learning how the body and mind work and making changes based on that learning. It takes time and patience and a certain level of commitment.

Brain wave biofeedback, or neurofeedback, is a different creature from standard relaxation training. Because the brain governs the whole body, brain wave training is considered much more global than other modalities that operate further "downstream." Neurofeedback is different than other kinds of biofeedback in another sense. Muscle biofeedback, or EMG, for example, requires a fair amount of conscious effort. A client must be aware of his or her clenched jaw muscles during the session, but also afterward. EMG clients must consciously work at letting go of tension after they leave the therapist's office and often wear a blue dot on their hand so they remember to change their habitual tension and relax certain muscles or muscle groups. Brain wave training is much more automatic. A person does a forty-five minute session. After they leave, it's over. Nothing to be mindful of, no blue dots. Even though a client is changing his or her own brain, the training is so powerful and simple that it is very different from other kinds of biofeedback.

After Joe Kamiya's work with Richard Bach, early reports by neurofeedback proponents indicated that alpha training might be a panacea that would reduce all of the body's accumulated stress all at once, a healing system that would normalize all body function. By reducing stress in the brain, which controls everything else, muscle tone, hand temperature, and skin conductivity would be brought to a healthy, comfortable state. And it would alleviate stress in the tissue of the brain as well. Brain wave training would clear the body and the mind.

Kamiya's work generated a great deal of excitement in the psychological community. One evening, he mentioned his work to Abraham Maslow, the founder of the humanistic school of psychology, and the next morning at six A.M. Maslow called him and said brain wave training implied an undreamt-of level of control that was so exciting he had been unable to sleep. Other researchers heard of Kamiya's work and started research of their own. *Psychology Today* did a piece on Kamiya in 1968, and things really took off. Some people who came out of the alpha state reported feeling rested and sharp, as if they had awakened from a nap with a clear head, while others reported overwhelming feelings of tranquillity and reverie. Some of Kamiya's subjects clamored for more training, so they could experience alpha again. Some artists and musicians reported experiencing intense rushes of creativity, and others described sensations of floating. Yet some people reported nothing. "We found that people who were relaxed, comfortable, and cooperative tended to produce more alpha waves than those who felt tense, suspicious, and fearful or who actively thought of what was going to happen next," Kamiya says. He also says that he believes the production of alpha depended a great deal on whether there was a rapport between the researcher and the subject. "It's a key factor," he says.

Despite the variety of experiences, word got out that scientists had stumbled onto a wonderful and halcyon state of being that everyone could access, a Shangri-La within the mind. "Instead of gulping a tranquilizer," Kamiya said in *Psychology Today* in 1968, "one might merely reproduce the state of tranquillity that he learned by the kind of training used in our studies." The popular press loved the story—it seemed the future was here—and spread the word. "Joe Kamiya has become something of a pop hero to kids who hope to groove their way into an instant satori," said one writer. "Though the exploration of autonomic control is still in its infancy," said *Time* magazine in a story about Kamiya, "the vistas it opens are staggering." "The children of the future may look back on us as little more than Neanderthal men, crude creatures who were unable to control our feelings, our physiology—and unable to play upon the instrument of the brain," said *The New York Times Magazine*.

Indulging in a little showmanship, Barbara Brown, a top biofeedback researcher and the author of several popular books, hooked up an electric train in the lab for reporters. When a subject generated alpha, the train would chug down the tracks; when the alpha ceased, the train stopped.

Barry Sterman was training cats as alpha training was making its debut, doing his higher-frequency work. The difference between alpha and beta training is more than just the difference in frequencies. They are distinctly different approaches. Beta training, which includes SMR, is more direct and physiological. It is very site specific; it directly treats the neurons that govern different aspects of motor abilities and senses, and exercises those areas, strengthens them, which enhances the abilities and senses they govern. Alpha and other deep states are more generalized approaches and more closely resemble the psychotherapy model. They guide the body and mind into a frequency-specific, profoundly relaxed state, which releases stress in the body and in the brain. The two approaches are used to treat many of the same problems from different perspectives.

By the late sixties, laboratory research into brain wave training was blossoming. The first meeting of biofeedback professionals took place at the Snowmass Resort in Aspen, Colorado, in 1968 as part of the International Brain and Behavior Conference and included the leading figures of the movement, among them physiologist Barbara Brown, Joe Kamiya, and the chairman of the meeting, Les Fehmi. In October of the following year the first regular meeting of biofeedback researchers was held, at the Surfriders Hotel in Santa Monica, California. With 142 attendees, this meeting included the major researchers, including Thomas Budzynski, an expert on rapid learning in the hypnogogic, or "twilight," state; sleep expert Johann Stoyva; Elmer Green of the Menninger Clinic; Barry Sterman; and Thomas Mulholland. Barbara Brown was elected the first president. A critical mass for a new and exciting field was building, and the enthusiasm was palpable.

By then the word was also out to the general public, and the meeting was packed. "It was a mixture of uptight scientific types of all types, and people barefooted, wearing white robes, with long hair," says Kamiya. "It attracted the heads to a tremendous extent." This was the time of Timothy Leary, Ken Kesey and the Merry Pranksters, free love, LSD, the turned-on Beatles, and the Maharishi Mahesh Yogi of transcendental meditation fame. There was a strong and pervasive belief that altered states of consciousness were a panacea for the "uptight" society that was waging a vicious war on Vietnam, despoiling the earth with toxic wastes and worshiping profit and greed. While some researchers in the field had similar feelings about the promise of biofeedback, the emotional, often wildly speculative claims made others, like Sterman, extremely uncomfortable. "Half of the group had ponytails and saffron robes, and half had crew cuts and ties," says Sterman. "We'd look at each other and say, 'What planet are you from?'"

It was at this first meeting that the pioneers got together to give their nascent field a name. Someone proposed the Society for Autoregulation, but people complained that it made them sound like car mechanics—one person joked that they might get sued by the Teamsters. The Feedback Society was offered up. The Society for the Study of Consciousness. "I favored self-regulation," says Sterman. But Barbara Brown's proposal of biofeedback—*bio* meaning life and *feedback* meaning information returned to the user—won the vote. The group became the Biofeedback Research Society, which later changed its name to the Biofeedback Society of America and then to the Association for Applied Psychophysiology and Biofeedback, which it is called today. It is the largest biofeedback organization in the world, with more than two thousand members.

Other researchers began to push the envelope on alpha brain wave research, and the work was controversial. One of these was Elmer Green, now eighty-one, who is probably the most famous of all biofeedback researchers. Green grew up in a Minnesota home where the parapsychological was an accepted part of life, and his research incorporated that acceptance. In his best-selling book *Beyond Biofeedback,* he wrote of a vivid dream he had about a friend's brother, a sailor, who complained of an infection in his leg. Green woke with a start at the intensity of the dream but thought no more about it. A week later, the friend told him that his brother was in a navy hospital with a severely infected leg.

Green steeped himself in psychic teachings in the 1930s. In 1939 he met his wife, Alyce. They married in 1941, and Green received his undergraduate degree in 1942, in physics. With four children to support, he took a job at the Naval Ordnance Test Station, a government facility in the remote desert of Southern California where, as a specialist in all types of instrumentation, he headed a team that researched and designed guided missiles and rockets. In some ways, rocket science was, in fact, a kind of metaphor for biofeedback, Green wrote later. A heat-seeking missile called a Sidewinder, for example, zeroes in on its target by detecting heat from engine exhaust. It transmits the information to an on-board computer, which constantly adjusts the missile's controls.

In 1962, while Green was studying at the University of Chicago, someone gave him a copy of a book called *Autogenic Training* by Johannes Schultz, a German psychiatrist and neurologist. The book discussed methods of reducing stress through relaxation and self-hypnosis. The work struck a chord with Green, who was studying the relationship between mind and body. In 1964 he and Alyce moved to the Menninger Clinic in Topeka, Kansas, to establish the Voluntary Controls Program, a pioneering program in biofeedback. In 1965 he attended a biofeedback conference and heard from Joe Kamiya about his work. Green added that to the other tools he was using at Menninger.

In 1973 Green, who worked with his wife, Alyce, undertook one of many projects that would bring a great deal of attention to the field of biofeedback when he and a research team took a portable psychophysiological lab to India. They traveled more than seven thousand miles across the subcontinent to study the Eastern holy men and their mastery of physiology, using the tools and methods of Western science. One of the most intriguing experiments took place at the southern city of Pondicherry, where the subject was Yogiraja Vaidyaraja, the so-called "burying yogi," who often spent two or three days buried in a box several feet underground to demonstrate his devotion to his followers. For the purposes of Green's test, the yogi was sealed in the lotus position in a completely airtight cube, 3½ by 3½ by 5 feet, above ground. The crate was made of wood and waxed on the inside to prevent the flow of air through the wood. One panel of the crate was a door, made of quarter-inch plate glass, through which Green and his team could see the yogi. The cracks around the door were sealed with a quarter inch of polyurethane foam. "I was convinced," wrote Green in *Beyond Biofeedback,* "that the box was sealed more tightly than the average refrigerator." The day before the experiment, an associate of the yogi had burned a candle in the box, and it had gone out, for lack of oxygen, after about an hour and a half.

On the day of the experiment, the yogi was led to the box and seated himself in the lotus position. Numerous wires were attached to the man to measure, among other things, his heart rate, his galvanic skin response, and his EEG. A physician, Green, and others in the experiment estimated how long the yogi would last. Green thought the yogi would reach his limit in two hours, while the doctor said he would need to be let out after four hours to avoid unconsciousness. After nearly eight hours in the airless box, the yogi signaled his desire to be let out, complaining that he had received three electric shocks from the equipment. The testers were flabbergasted. During his stay, readouts showed that his breath rate had dropped to less than four breaths per minute, and his pulse rate had dropped by more than half. After the yogi emerged from the box, researchers tried to take a blood sample, and they couldn't—the holy man had withdrawn the blood from his extremities. The EEG reading showed that the yogi had "produced alpha almost continuously," Green wrote. Green also found that the yogi had the unusual ability to move quickly between disparate brain frequencies. When he entered the box, he moved nimbly and almost instantly from beta, or waking consciousness, into a deep alpha state but did not slip into theta, the beginning stage of sleep. This flexibility is a hallmark of all the unusual subjects and holy men whom Green studied over the years.

One of the former was Jack Schwarz, a man with an astounding ability to control his physiology every bit as effectively as a yogi; he could punch a needle completely through his arms and legs without blood or pain. He was hooked up to an EEG monitor at Green's laboratory in Topeka, and it showed that as Schwarz began piercing his arms, his brain waves dropped suddenly and deeply into alpha.

This kind of work intrigued the public as much as it alienated scientists. The scientific culture is a conservative one, and most of its members prefer to keep an arm's length from popular trends. Science demands rigorous, long-term testing before claims can be made. For the mainstream researcher, wild claims being made about alpha biofeedback were appalling and irresponsible.

There were other problems with the claims being made for alpha. Other researchers tried to replicate some of the claims in the lab and couldn't because it's a difficult thing to produce the frequency correctly in laboratory conditions. Companies marketed machines to create instant alpha, but the equipment of the time was primitive, often unreliable, and difficult to use appropriately. Moreover, alpha on one site of the brain is not always enough to produce deep feelings of relaxation, let alone enlightenment. Alpha brain wave training did not live up to the hype, and many scientists dismissed the mystical alpha state as so much New Age folderol.

At the time, researchers were exploring another area of brain wave biofeedback, theta training; but only a handful of researchers explored it, and it received scant attention. Theta waves are very low-frequency waves, from 4 to 7 hertz, below the alpha range, but before the onset of sleep's delta waves. The dreamy state is difficult to study, because theta waves are hard to sustain without falling asleep, even with biofeedback. Yet it has intriguing properties—the Greens studied it and said theta was associated "with a deeply internalized state and with a quieting of body, emotions and thoughts, thus allowing 'unheard or unseen things' to come to the consciousness in the form of hypnogogic memories." Theta is apparently where childhood memories are "stored," and people who tap into this area often experience the emergence of long-buried events from their past. The images are not foggy ones; they are often a vivid reexperiencing of things. Researchers also found that spending time in theta had a tranquilizing effect on clients that lasted two or three days.

Theta also appears to be associated with the uncritical acceptance of new material, and researchers have found that people can learn things much faster if the material is presented to them in this state. Thomas Budzynski, now a neurofeedback researcher at the University of Washington, developed a device he called a Twilight Learner. Tape-recorded messages switch on as the frequency reaches theta, and the information that people want to learn plays through headphones. The message gets louder as the frequency diminishes so the person doesn't fall asleep. "People can learn a lot of material very quickly," says Budzynski. Theta is also a state of hypersuggestibility. Some subjects had suggestions "implanted" in this state and felt that it made their efforts to do such things as stop smoking and lose weight much more effective.

Partly to blame for the marginalization of biofeedback, says Elmer Green, is a strong bias among scientists against the study of human consciousness. Science yearns for measurement, and the changes biofeedback makes are largely subjective. He recalled a meeting with a site review committee from the National Institutes of Mental Health, as its members considered funding a grant at Menninger. Green brought up the subject of Johannes Schultz's autogenic training, and the group, he said, "looked at the floor, shuffled their feet, and coughed." After he mentioned something about human consciousness, one of the men said, "No one knows what consciousness is, but we can detect its absence." Then, Green recalls, the group broke into laughter.

As research funds dried up in the 1970s, the study of altered states dwindled. But a handful of scientists remained in the field, shifting from pure research to clinical practice as a way to make a living while continuing their work. Old-guard alpha guys like Les Fehmi, who still does alpha training at his office in Princeton, New Jersey, believes that the glimpse in the 1960s of what alpha could do was very real, but the hype came along before the field was mature, and the research was cut short. Now, he says, with the advent of computers and three more decades of research, alpha training has come into its own as a powerful way to deal with the myriad problems caused by stress.

Sterman's higher-range work, meanwhile, did not become popular in the 1970s as alpha and theta training seized the imaginations of many. Instead, it nearly died out, and it might have disappeared completely if not for the work of a handful of determined clinicians, and one remarkable and controversial woman in particular.

CHAPTER FOUR

Lazarus?

∞

"I want my son back," Marvin Ritchie said to Margaret Ayers, who stood over the wan, pale figure of twenty-year-old Jay Ritchie as he sat motionless in his wheelchair with his chin heavy on his chest, oblivious to the world around him. Ayers, who had once worked with Barry Sterman at Sepulveda, had left his laboratory, opened a neurofeedback clinic, and expanded the technique far beyond epilepsy in her solo practice. People sought her out for the astounding, one-of-a-kind service she claimed she could provide.

On February 11, 1996, the energetic Jay had just graduated from college and was working in an auto body shop in Hastings, Nebraska. He had a passion for anything with a motor and was an avid racer of "micro midgets," a type of very small racing car. He was working beneath a 1974 Roadrunner that was up on blocks, and when he tried to pull a shock absorber loose, the car came off the supports and the corner of the frame slammed into his chest, landing just below his ribs. Then the vehicle slid off of him. The blow had stopped his heart instantly and left him unconscious. A few minutes later, Jay's boss came out, pulled him from under the car, and called an ambulance. The paramedics treated him on the way to the hospital, but according to his mother, Deb, by the time his heart was beating again, he had been medically dead for at least seven minutes, perhaps longer.

Jay suffered something called anoxia, a lack of oxygen to the brain that resulted in what his mother said doctors diagnosed as a coma, a Level II, the second-most severe on the Rancho Los Amigos coma scale. His mother said he couldn't be roused by light or sound, and if ice was dropped on his bare skin, he did not react. After spending more than a year in three different hospitals, Jay was brought to his home in Ansley, a tiny town on the grid of rural highways in south central Nebraska, sixty miles northwest of Grand Island. His weight had dropped from 135 to 96 pounds. "He would open his eyes sometimes, but he wasn't really there," said his mother. "Doctors more or less told us that he would never improve from the state he was in." There was no interaction between Jay and his family.

Deb and her husband, Marvin, who both work on an assembly line for a company that makes test tubes and other medical equipment, read an article in a local paper about another young Nebraska man who had been treated with an unusual technique called neurofeedback, which had brought him out of a coma. The woman who did the work was named Margaret Ayers, and her office was in Beverly Hills, California. The Ritchies called Ayers and arranged to travel to Beverly Hills for a series of treatments.

The Ritchies had never flown on an airplane, but it was something that Jay had always wanted to do. His first time aloft, however, he sat with his head down throughout the trip. The family checked into a hotel a few miles from Ayers's office, and the next morning the Ritchies took Jay there for his first treatment. Ayers hooked him to her computerized biofeedback program, which like all neurofeedback takes the faint EEG signals—she calls them "whispers from the brain"—and amplifies them. She also connected him to something she calls a "coma box," a small, black, metal unit with a green light, which is her design. It sat about four inches in front of his head. As Ayers held one eye open with her fingers, the green light, exactly the same size as a dilated human pupil, was aimed at Jay's eye. She began talking to the boy as if he were awake, above the hum of the respirator, coaxing him to experiment with changes in his state of mind until the green light came on and shone in his eye. "I want you to come out of the coma, Jay," Ayers coaxed, talking to the comatose boy in a soothing tone of voice. "Your parents love you very much, and they want you to come out of the coma. They've spent a lot of money to bring you here, and they want you to get well. If we can get you out of the coma, we can make you freer." In just a few minutes the small green light blinked on signifying that the boy was inhibiting his theta waves. "Excellent, excellent," Ayers coached in a friendly tone of voice. "Good, good, good. Get that EEG down."

Jay had no idea what a theta wave was. He was apparently awake inside his still body, however, and could understand what Ayers told him as he lay there: You need to make the green light shine into your eye. He did not know when he was inhibiting theta, as Sterman's cats and Mary Fairbanks did not know they were producing SMR. They were all working for the reward. The equipment and the operator did the rest.

Jay Ritchie had to inhibit theta because the low-frequency waves had taken over his brain. In a brain that is functioning normally, theta waves are dominant only when a person is drowsy or on the edge of sleep, the dreamy place called the hypnogogic state. When a brain is damaged, by a blow or, like Jay's, from a lack of oxygen, the brain's regulatory apparatus is thrown out of whack. Theta waves spread across the brain, and beta waves, the higher frequencies, which construct our waking consciousness and our senses and motor skills, are subsumed by the lower frequency. Comas, says Ayers, are at the most severe end of "too much theta," while head injury falls somewhere in the middle, and attention deficit disorder is at the least severe end of the spectrum.

Ayers talked to Jay for twelve minutes and then took a break to allow him to rest, after which she treated him for another twelve minutes, in a similar style. "I heard you are a good mechanic, Jay," she said. "I understand you used to love to race cars with your father. Wouldn't it be great to get behind the wheel again? Make the light come on, Jay, and perhaps that will happen."

The training seemed to rouse Jay a little, right away. "The first session seemed to change him," Deb says. "He picked up his head and looked around." And later that day, back at the hotel, Jay Ritchie said his first word since he had been plunged into a coma—"Mom."

Every day for a week Ayers conducted sessions. And at the end of the week, Jay Ritchie was a very different young man. "On the plane on the way back he picked his head up and looked out the window," Deb says.

Since then, Jay has had dozens of sessions, and Deb and Marvin say they continue to gain more of their son back with each treatment, far more than they ever expected. "He never focused on anything," Deb says. "His skin didn't feel like skin—he looked like a person with plastic skin. He was just a body. He's more Jay now. His skin is pink and normal," she says. "They had him on sixteen different kinds of medication, and then we reduced that to four. He doesn't take any now. We see a little more of him come out every day. He's real awake and alert right after a session, and you can really see him trying to sort out what you just told him. He can speak a few words—'Mom,' 'no,' 'yum-yum,' 'Amy'—that's his niece—and 'Marvin,' his dad. He responds to things, he turns his head and moves his eyes. Especially racing motors. The instant he hears them he turns his head. We raise huskies and sell them. We had seventeen puppies this year, and we bring them in and they lay with him. He gets kind of irritated when they lick his face. Oh, and the cats, he really responds to the cats. They were his cats, and he'll watch them, and he'll try to move his hands to pet them. He has emotions too. If we see tears in his eyes, emotion tears, we have to ask him, 'What's wrong? Are you sad? Do you have a headache?' and he'll answer yes or no by blinking his eyes—yes is once, and no is twice."

"I've seen tremendous changes," says Jerry Denton, a psychologist who runs the Prairie Institute of Psychology in Grand Island, where he has continued the treatment regimen that Ayers began, using her equipment and protocol with Jay. "He's completely out of any stage of the coma. He's working on speech. He's in physical therapy. He has some control over his hands. You can see the brain waves changing through time in his EEG, and it's dramatic. I think he's going to make a lot more progress. In two or three years he'll be able to move about and speak some limited phrases."

It all sounds too good to be true, a miracle of some sort. But it is neither of those, Ayers says. There is science behind it, the science that Sterman and other brain wave trainers began. Ayers claims that she has brought thirty people out of the twilight world of coma, including one man in Texas who now leads the same life he had before he was injured. Some of the coma cases have been videotaped as they regain their ability to engage with the world.

The office where Ayers treated Jay Ritchie and hundreds of other people is on Cañon Drive, in a tony, sun-washed neighborhood amid the palm trees and high-end retail stores in Beverly Hills. It is a warren of tiny rooms on the second floor of a small, informal office building called the Courtyard Building, which was built by Will Rogers in the 1920s to house his office. The courtyard is full of greenery and is ringed by shops that sell expensive gifts and clothes and beauty makeovers. In addition to her clinical business, called Psychoneurophysiology, she sells her instruments out of the office, a business known as Neuropathways EEG Imaging. In her fifties, Ayers is trim, with short hair. She is businesslike but warm and was dressed in a beige pantsuit and a red blouse the day we met. Though she is polite, as talk turns to the history of neurofeedback and the way she says some of her colleagues have treated her, there is a strong undercurrent of bitterness. Over lunch at a bustling Chinese restaurant down the street from her office, Ayers told me how she became one of the country's first clinical practitioners of neurofeedback.

Ayers was born in Albuquerque and raised in Seattle. She earned a bachelor's degree in microbiology, a master's degree in psychology, and did four years of doctoral work in clinical neuropsychology. She never wrote her dissertation and so never received a Ph.D. In the early 1970s one of her professors, who had mentored Barry Sterman, called Sterman to ask if he would allow her to work in the lab at Sepulveda doing EEG research for her doctoral thesis, as part of her studies with the UCLA Medical School. Sterman agreed, and Ayers moved to Los Angeles from San Diego to work with him, learning to do EEGs on epileptics as part of Sterman's project.

Early on in Ayers's two-year tenure with Sterman, as she did the EEG study, Sterman asked Ayers to assist him with neurofeedback protocol on a young quadriplegic woman from a well-to-do South American family. It was her first work with biofeedback. "When we were done, this individual was able to feed herself," says Ayers. For a young graduate student, she recalls, "it was very exciting." She and Sterman wrote up the work, and it was published in a biofeedback journal in 1976.

Ayers remembers that, as she treated the young woman, and as Sterman and his assistants worked with epileptics, she noticed a profound change wash over these chronically ill people. It affected not only their physical condition, but their mood and demeanor as well. "These epileptic individuals were happier, smiling, they were talking about things," she says. "I thought to myself, 'You know, the sensory cortex is next to the motor cortex in the brain. We may be affecting emotions.'" Ayers had been captivated by the success of her first work with neurofeedback. As part of her own research, she had also been studying the electroencephalogram of a kind of depression known as genetic unipolar depression. "I asked permission to look at the all-night sleep of genetic unipolar depressives," she says. Depressives, Ayers found, do have a unique EEG, which shows in a disorganized, atypical sleep pattern. People with normal sleep patterns pass through stages 1, 2, 3, and 4 of sleep and then bottom out in the deepest state of REM, or rapid-eye-movement, sleep, where dreaming takes place. "Depressives go one, two, three, and REM," says Ayers. They were missing part of their deep sleep levels, she realized. "I also noticed that they had delta and theta waves, which are sleep waves, in their wake state. So sleep frequencies were invading the wake state, and the wake state was invading the sleep state."

In 1975 Ayers left Sterman's lab and set up shop in an office on Wilshire Boulevard with neurofeedback equipment that she leased from the family of Sterman's technical person, Sid Ross. It was an archaic piece of equipment compared to the computerized units in use today, and Ayers has a museum piece called the Neuroanalyzer 4000 Electroencephalograph displayed in her office. It has two lights on it: a red one, which blinks on when a person is producing too much theta, and a gold one, which comes on when the person is appropriately producing beta. There is also a counter that adds up the number of times the person gets the gold light to come on, providing another level of reward.

To test her protocol, Ayers trained fifteen depressives to inhibit, or discourage, slow waves and keep the sleepiness out of their wake states. "These individuals had all been hospitalized once; they'd had drugs, they'd had shock therapy," she says. After the training, "they all reported feeling better, having more energy." Ayers hung out her shingle as one of perhaps two or three people in the world treating with neurofeedback. No one at the time was taking on the kinds of cases she eventually did, which included everything from epilepsy to closed head injury to stroke to comas to attention deficit disorder. "It was so exciting," she says.

She has even used the machine on herself. "I had a very bad hand familial tremor," she says, referring to a malady in which a person's hand sometimes shakes uncontrollably. "I would give lectures, and psychologists would say, 'That was a great lecture, but you really ought to do something about your anxiety.' And I would get very angry and have to explain that it was genetic. One day I decided to hook myself up on the hand in the motor area of the brain to see if I could get rid of it. And I got rid of it." She held her hand out over the restaurant table. "You can only barely notice it. You can only notice it if I have coffee in the morning and no food."

Ayers claims that her success is due to two approaches that are unique in the field of neurofeedback. For closed head injuries, coma, and stroke, too much theta is usually at the root of the problem. It keeps the brain from getting into higher frequencies, which allow the conscious mind to function normally. Virtually every other neurofeedback practitioner encourages beta and SMR, and many discourage, or inhibit, theta at the same time. Ayers says she primarily discourages theta, because the central nervous system is inclined to inhibit, rather than encourage, impulses and frequencies, and so inhibiting theta is a more natural process for someone to learn.

Ayers also believes her therapy works so well because of her neurofeedback instrument, which she claims is a one-of-a-kind design that makes it more powerful than any other instrument. Most instruments, she says, take the EEG from the person's brain, average it, and reconstruct a signal based on the average. There is a delay, she believes, between the real signal and the time it is displayed on the screen of three tenths of a second to a full second. Because the human nervous system works in milliseconds, she says, the split-second delay is critical. "The event happened too long ago to be connected to the feedback, and you end up giving positive feedback for abnormal EEG patterns, causing an increase in those abnormal patterns," she says. Her instrument, which she designed with the help of an engineer, is much faster. With less than a millisecond delay, she claims, her machine displays the information virtually as fast as the nervous system can process it. That means the brain is reinforcing the actual event. Ayers says that's why she often gets results in one or two sessions.

Ayers seems to be the only one in the industry who thinks her machine does something others don't. Other manufacturers say that

Ayers's claim to superior equipment is specious. Experts I talked with say that all of the different makes of neurofeedback equipment have a delay in the tens of milliseconds to allow the machine time to process the data. Such a delay is not significant to the brain, and if Ayers were feeding back one-millisecond information, it would be overkill. The brain doesn't work that fast. But they don't think it's technically possible for Ayers to be feeding information to the user that quickly. She would have to be feeding the raw signal onto the screen, which would be so energetic it would flicker and not be watchable. It has to be smoothed. And that takes at least tens of milliseconds to do, which would make it comparable to other manufacturers' instruments.

By successfully banishing the unwanted invasion of theta into the waking state, Ayers says she has taken on, and successfully treated, many of the most difficult cases in the medical world: drug addiction, alcoholics, head injury, stroke, the symptoms of cerebral palsy, Parkinson's disease, depression, all types of seizure disorders, attention deficit disorder and hyperactivity, visual dyslexia, auditory dyslexia, alexia, aphasia (a loss of speech), and a wide variety of other learning disabilities.

One of Ayers's approaches is with a pinpoint electrode placement. When a client suffers from aphasia, for example, she will place the sensing electrode on Broca's and Wernicke's areas—the tiny patches of gray matter in the cortex that govern speech—and train that part of the brain to inhibit slow waves. For problems with vision, she trains the occipital lobe, which governs eyesight; if someone has had a stroke and has a paralyzed leg, she trains the site on the sensory cortex that governs that leg. "I've seen people walk better in one treatment," she says.

The most vital contribution neurofeedback has made, Ayers believes, is in the treatment of head injury. Head injuries, or traumatic brain injuries, have a variety of causes, from bashing into a windshield in an auto accident to falling on icy pavement. Brain injuries can also be caused by a heart attack that cuts off oxygen to the brain, as was the case with Jay Ritchie. "The number of patients with head injury in our country exceeds heart attacks and disease," Ayers says. "People with head injuries look normal, but because they don't drool or they're not in a wheelchair, it's not labeled as a disease. Yet it's a serious problem. These people can't work. They can't drive a car because they can't remember the sequence of events. They can't watch a movie because when they get to the end, they can't remember the first thirty minutes."

David Pera, a motorcycle racer from Temecula, California, is one of those people. In a desert race between Reno and Las Vegas, his bike had suffered a loss of power going up hills. He took it to the shop that had worked on it, and the mechanics there asked him to demonstrate the problem. He said he couldn't do it at the shop. "There were two problems," he says. "There was no hill, and I didn't have my helmet." The mechanic told him to do a wheelie, or ride on the back tire, which would replicate a hill climb. Don't worry about a helmet, he said. Pera drove down the alley in back of the shop and did a wheelie. As he came out of the alley, a catering van blew a stop sign and smashed into him. "I was in a coma for seventy-seven days," he says. "I died twice, and they had to perform two emergency operations." When he recovered, he says, "I couldn't remember anything." His short-term memory was gone, and anything that people told him disappeared within seconds. He soon gave up his business and sank into a depression. Then a friend mentioned Ayers. After a single session, he says, his memory started returning. Four months later, with his theta inhibited for an hour once a week, his memory is back. "It's Margaret," he says. "I'm behind her a hundred and ten percent."

Ayers claims that her success rate with head injuries is 95 percent; with seizure disorders, about 85 to 90 percent; and ADD and ADHD, about 90 percent. An independent evaluation of the cases Ayers has worked on would be difficult because she has published little of her data. She has published several papers, though not in mainstream scientific publications. In one, a collection of the results of treatment of 250 patients with closed-head injuries, published in the *National Head Injury Syllabus* in 1987, she claimed that all her patients experienced an increase in energy and short-term memory and a decrease in depression, temper, and headaches. Unlike the work that Sterman did, her results are clinical and not based on careful, long-term, double-blind, controlled laboratory research projects. As powerful as the testimonials are, they are what scientists call anecdotal reports, accounts by people who have benefited from treatments in a clinical setting. Critics argue that such accounts are self-selecting; the people who benefit tell their stories, while those who see no, or little, gain go on to other things and are forgotten about. And there is the often heard claim that all the neurofeedback practitioner is doing is engaging the placebo effect.

Ayers vehemently disagrees that her work is nothing more than placebo and says critics have not taken the time to examine the cases. "If you bring someone out of a coma in two treatments who has been in a coma for two years from anoxia, and they say some words and they don't go back to what they were; or a stroke patient who you don't get until four years post, and their arm is tight against their chest and their nail is digging in the hand, and they're thinking about cutting the tendons and the patient has tried physical therapy, occupational therapy, hyperbaric oxygen, and acupuncture and [after neurofeedback] the arm is down and they are painting with it and it stays down, you know it's not placebo."

There are many other people treating mild closed head injuries with neurofeedback, most of them using equipment and approaches that differ from what Ayers uses but getting similar results. There have also been several studies that show the efficacy of the approach. Dr. Jonathan Walker is a neurologist who runs a clinic called the Neuroscience Center in Dallas, Texas. In the past decade he's treated more than 200 head injury patients with neurofeedback. In the past four years he's treated one hundred or so head injury patients with neurofeedback. "Most of the time it improves symptoms or completely eliminates the problem," he says. "There's no effective drug treatment, no effective treatment at all, except biofeedback. And the brain stays recovered." In a head injury, Walker says, the axon, which is part of the neuron, fractures and can no longer transmit information. His model for what is happening is that the brain learns to work around the damage. "There's a lot of redundancy in the brain," he says. "And other cells near the damaged one can be trained to take over the function. Stimulating nearby neurons strengthens existing pathways."

Some scientists and medical specialists are not convinced. If Ayers is able to do what she claims, it is nothing short of a revolution in care for brain-injured patients, says Dr. Ronald Cranford, a neurologist in Minneapolis who specializes in coma and other brain trauma and was head of a Coma Task Force in the early 1990s that looked at research literature regarding alternative therapies for vegetative states. The report of the task force was published in the *New England Journal of Medicine* in 1994. The task force did not consider brain wave training for coma, but they studied reports from a range of other alternative treatments, from hyperbaric oxygen to sensory stimulation. "We didn't feel there was anything that conclusively showed any significant improvement," he says. He also says that the use of unproven cures is growing as people with serious head injuries survive longer because of better medical care. The task force did not conduct studies of the alternatives; it only reviewed existing data. Even though some approaches may work, Cranford says, the cost of scientifically testing them may be prohibitive. "We estimate a study would cost ten million dollars," he says. "You would need at least one hundred patients and need to make a judgment that these patients would be treated for at least two years. It would be very expensive, very time-consuming, very ethically problematic. It's loaded with huge, insurmountable challenges." So neurofeedback treatment for coma may never get thoroughly studied.

Chris Carroll, the psychologist mentioned earlier, knows firsthand about the power of the technique—he has used it to treat ADD and ADHD children. He believes the work that Ayers does may be within the realm of possibility, and he wants to believe that Ayers is doing what she says she is. But thorough studies have not been done, and he worries that claims unsubstantiated by controlled studies similar to those conducted by Sterman will hurt the reputation of the field among scientists. Extraordinary claims, it's been said, require extraordinary evidence. And for Carroll that evidence isn't in yet. "All science begins with anecdote," says Carroll. "What Ayers has done is clear that first hurdle in the scientific process. Creativity needs to be encouraged to explore the possibilities that apply in any area—medicine, psychotherapy, every field. As long as we do no harm. That should never be violated. The next step, however, is to use the most appropriate available scientific methodology so we know how consistent her results are. It may be true but not very consistent. That's the responsibility of science. She needs to treat people who have been randomly assigned to her, for example. Other people need to be able to replicate her work. She could be engaging a mechanism for healing in these people that has nothing to do with neurofeedback." The green light could be stimulating the patient, he says, or all of the attention could be responsible.

Carroll thinks there is a simple, inexpensive way to test it. "A controlled study that would randomly assign coma patients to the Ayers protocol or a control group would separate out a lot of confounding factors," said Carroll. "That could easily be done in a double-blind format with all of the patients receiving the same treatment, but for the control group, the small green light would come on in a random fashion. If patients were randomly assigned to the two groups and the group for whom the green light was contingent upon decreasing theta activity showed significant improvement as rated by impartial, blind observers, her conclusions would carry a lot more weight." Ayers's claims of bringing people out of comas are so extreme, even in the field of neurofeedback, and the research is so thin, that her work will be dismissed until some kind of study has been performed. Ayers has not taught others to perform the treatment protocol for coma, which also casts doubt on its efficacy.

Carroll also disagrees with Ayers that the fact that a paralyzed hand relaxes after training, when other modalities have failed, is proof positive of the efficacy of the technique. "There are many ways of engaging the influence of mind-body connection," he said. "Just the fact that she is paying attention to someone may make a change. She really needs to prove that neurofeedback is the active ingredient of what she is doing."

On the other hand, some argue, there are numerous therapies, radical medical and psychological therapies, administered by top health care professionals every day that are not supported by careful studies. Despite the lack of studies, they are used because they seem to work.

The other factor to consider is the cost-benefit analysis of neurofeedback. If you have a loved one in a coma or who suffers from a closed head injury, and other therapies—which are few and largely ineffective—have had negligible impact or have failed, why not try something that hundreds if not thousands of people testify is very real? Especially something that is so inexpensive. Ayers charges $45 a session. As long as the patient knows the technique is experimental, why wait years for studies to be done, if they ever will get done? There is a huge gap between what the researchers are studying and the needs of clinicians. Even those who want to see more research conducted will come to Ayers, or another practitioner, if faced with the right situation. "If I had a loved one who was in a coma, you bet I would bring them to see her," says Carroll. "The right studies might matter to scientists, but they don't matter to families."

Science is a funny thing. The popular image is one of thoughtful, white-coated scientists working in a lab in a methodical way, doing their research, taking into account and building on the work of others, publishing their work, and participating in the give and take of scientific debate. The world of scientific research, however, can be a strange, dark, even twisted, place. In nearly every scientific field, the landscape is covered with a tangled thicket of ego clashes, vitriolic attacks, slights and perceived slights, accusations and counteraccusations. Though the personal disagreements play out in a kind of proxy war—the scientists usually attack one another's science and methods—there is a great deal of overlap between legitimate issues and personality clashes, and it is nearly impossible for an outsider to tell where one ends and the other begins. Though the science is what's most important, personality conflicts are sometimes inseparable from the story. That is very much the case with neurofeedback. There is a great deal of bad blood between some of the major players in this burgeoning field, and it is difficult to sort through all of the intrigue, some of which has played out over decades. This is a part of the reason there have not been more studies of neurofeedback and it has not become better known than it is.

Such is the case with Margaret Ayers and Barry Sterman. When Ayers left Sterman's lab at Sepulveda, Sterman says it was because she wasn't doing the work she had come to do. Ayers says she left because she was uncomfortable in an all-male club, and because she saw the incredible power of this new technique to help people. Here the plot thickens a little more: Sterman claims she was able to do this by secretly enlisting his full-time technician, Sid Ross, to lease her an unauthorized copy of the neurofeedback machine that Sterman and Ross, who is now deceased, had collaboratively designed. Unknown to him, Sterman says, Ross had patented the machine.

When alpha biofeedback became a laughingstock because of the claims some researchers were making, Sterman worried that without a great deal more research, using neurotherapy to treat patients would not only hurt his reputation but do further damage to the field of biofeedback in general as well. It was not ready to be used on a fee-for-service basis. Concerned that the instrument might make its way out of the lab and into the wrong hands, he says he allowed only qualified researchers to have one of the units. He says he required Ross to sign an agreement saying he would not sell it. To get around that, Sterman says, Ross leased the instrument to Ayers for her clinic. Sterman's nightmare was coming true. Incensed, Sterman says he tried to fire Ross, "but you can't fire a federal employee." The genie was out of the bottle. "The bad news is that Margaret did not know what she was doing," Sterman growls. "The good news is that it didn't matter." Ayers has gotten good results with her approach, Sterman grudgingly admits, which he says is a testament to the power of the neurotherapy, not Ayers's ability.

Ayers, for her part, is indignant at Sterman's claim that she acquired the instrument in any way that was less than ethical. When I told her that Sterman said she, in effect, leased a stolen machine, she laughed a hearty and cynical laugh. Ross developed the equipment she used, working on it in his own time, without input from Sterman. "It's a blatant lie," she says. "Sterman is just jealous. He never invented anything in his life."

Through the later 1970s and 1980s there was just a handful of neurofeedback practitioners. Ayers was treating full-time. Sterman was doing very part-time treatment of epileptics. Joel Lubar began his research with epilepsy and ADHD in 1974 and 1975, respectively. He then worked with Sterman for nine months in 1976 in order to observe Sterman's method for treatment of epilepsy. And there was Michael Tansey.

I attended a Tansey lecture to a roomful of neurofeedback technicians one fall day in Pennsylvania. His eyes bright, his goatee and hair a salt and pepper, he talked fast, excoriating other approaches to neurofeedback. Occasionally, as he emphatically made a point, or got angry as he talked about a subject that called up an old insult or he thinks is patently absurd, his voice would get shrill. Later, however, with a client in front of the machine, his demeanor softened. As he watched the subject's progress on a computer screen, he quietly and soothingly coached him to relax. "Floppy hands. Floppy. Floppy. That's it. Relax. Warm a stone in your hand. Floppy. Floppy."

Like Ayers, Tansey discovered Sterman's work and then built on those ideas to create his own approach, operating a neurofeedback clinic at an office in northern New Jersey until he retired from clinical practice in 1996 to pursue research. Tansey is temperamental, and discussion of his work is also tinged with bitterness, for he feels that his work has been not only ignored, but actively suppressed by Lubar, the Ph.D. psychologist at the University of Tennessee who pioneered work on neurofeedback and ADD and ADHD and who edits one of the field's key journals.

Tansey's approach to neurofeedback is very much his own. With a Ph.D. in child psychology from the California School of Professional Psychology in Fresno, Tansey began using biofeedback to treat patients for relaxation in 1976 but after a year became disenchanted. "I didn't see any therapeutic efficacy," he says. "It wasn't like Lourdes, where you see crutches hanging on the wall" after you treat people. Then he read about Sterman's and Lubar's work and began doing biofeedback in the higher ranges. Instead of using the 3-hertz SMR bandwidth, he put his own spin on it. He believes that each 1-hertz increment of frequency represents a different state of mind. While the idea is on the fringes of neuroscience, Tansey did publish the results of an intriguing study in a paper in the prestigious *International Journal of Neuroscience* in 1994, with his wife, Jennifer, and Ken H. Tachiki, the chief of clinical psychobiochemistry at the UCLA School of Medicine. They gave each of seventeen subjects verbal commands and then measured the way they responded—a kind of word association test that measured the frequency response in their brains instead of a verbal response. When subjects were asked to imagine a yellow ball, for example, the dominant hertz was 7, which led Tansey to associate 7 hertz with a mental image occupying a person's attention. When someone was given the one-word command "feel," 16 hertz was dominant, and Tansey associates that frequency with "nonspecific somatosensory awareness." Asked to add five plus six resulted in 10-hertz dominance—the frequency that governs "engaged thought." Tansey believes that one lone hertz is far more important than any other: 14 hertz. And he says it should be trained dead center, at the top of the skull, at what is known as the CZ site. In Tansey's experience, this is the only frequency most patients ever need, at the only site they will ever need. And it lies smack in the middle of Sterman's 12-to-15-hertz rhythm.

Tansey's model holds that people tend to get stuck in one frequency area or another, something he calls "parking places." The training helps push people out of their parking spot to teach the brain to function in other frequencies. Tansey has used what might be called the "perfect hertz" to treat everything from ADHD or hyperactivity, to Tourette's syndrome, to petit mal epilepsy; a variety of learning disabilities, including dyslexia; something called laryngeal dyskinesis, a problem caused by spasms in the respiratory system with symptoms similar to asthma; and more.

Tansey speculates that when some children are born, portions of their brains never quite fully unfold or form as the rest of the brain forms. Cells don't complete all of their connections, and key pathways that different parts of the brain need to create a network don't form. "Everything is there," Tansey says. "It just hasn't kicked in. It's like having a phone line through Budapest, and somehow there's this one neighborhood, this one circuit breaker that just doesn't kick in, and in that whole area you can't make a phone call. And if we work the system [with neurofeedback], those circuit breakers get tripped. All of a sudden this whole area, this whole matrix of stuff that was ready to unfold, starts unfolding. That doesn't mean that everybody who has a developmental delay is going to have an instant miracle and work their brain and all of these things are going to trip. But a lot of things do trip."

The CZ site on the sensorimotor strip, he says, governs volitional motor control. After a client does fifteen to thirty sessions, all of the areas of the brain that should have kicked in are operating, and the motor cortex strengthens and normalizes. "It's antimuscle tension," he says of his approach. "Every muscle everywhere. That doesn't mean you wet your pants. But all things that are in spasm smooth out over a couple of training sessions. So you get systemic motoric calm. Unstress. When you have that, you also reduce blood pressure and pressure on the entire vascular system. The muscle sheathing, that is, the muscle tension around veins and arteries, relaxes, and you get peripheral blood flow as a result of this; you nicely lower blood pressure. It's a very, very nice way of being mindful but not engaged. You see hyperactive kids walk in pasty-faced and white, they are so tight, and when they are done, they walk out pink." Tansey has carefully collected a good deal of data in his two decades of practice. He has published well-documented papers and has dozens of videotapes of his clients and the gains they have made.

In the early 1980s Tansey, Ayers, Sterman, and Lubar, and perhaps a couple of others, were the entire world of high-range brain wave training. They believed deeply in the power of what they were doing and felt that it was a revolution in health care. But they were also content to be clinicians, write a few papers, and wait until the world discovered them. Part of that patience came from a shibboleth of scientific culture: it is considered unseemly for scientists to popularize their work or to make appeals to laypeople. Another problem is that the concept of teaching people to train their brain waves was so foreign to any established idea of health care, and so difficult to describe, that many people simply would not listen or did not believe them. Had these few clinicians disappeared, or gotten out of the business, their ideas would have disappeared with them.

Neurofeedback, however, did not die out and is no longer the province of a few individuals. Early in 1985 a friend of Susan Othmer's gave her some cassettes of a lecture that Margaret Ayers had given. She listened to the tapes and was impressed with the alternative approach to epilepsy and the associated neurological problems that come with the disease. Sue and her husband, Siegfried, had a teenage son with temporal lobe epilepsy and severe personality problems associated with the seizures. They had come to alternative health care in the indirect way that many people do—when the road of mainstream medicine peters out, leaving them no place to turn and a serious problem still on their hands. Not only had doctors been unable to help their son much, the Othmers had found the medical industry to be a massive, uncaring, often offensive machine in which the feelings of a patient and his or her family were way down on the list of what was important. It was the old Cartesian approach: treat the body as equipment, and don't worry about the mind. The doctors had taken the blood samples, done the tests, written the available prescriptions, titrated the medication, and then thrown up their hands and said, There's nothing more we can do. But the Othmers wouldn't buy that. Their son Brian was sixteen, seriously troubled, both physically and emotionally, and faced with an uncertain future. So on March 5, 1985, a date permanently etched into the Othmer family history, Sue drove Brian from Sherman Oaks to Beverly Hills, where he sank into a chair in Margaret Ayers's office for his first session of EEG neurofeedback.

The Othmers could not have imagined what had been started on that spring day.

CHAPTER FIVE

Brian's Brain

∞

California was still the Golden State, the land of sun-soaked promise, when Siegfried and Sue Othmer moved to Sherman Oaks with their two-year-old son, Brian, in the fall of 1970. Then thirty, Siegfried was a young physicist with a new Ph.D., and he had taken a job at Northrop Research and Technology Center, researching ways to make semiconductors resistant to radiation in the event of a nuclear war, part of the booming Southern California defense industry. Sue was planning to continue the neurobiology studies that she had begun at Cornell, so she could earn her Ph.D.

Siegfried had met Sue FitzGerald at a party during their college years in New York. They married in 1964. In 1968, while they were doing their graduate work there, Sue became unexpectedly pregnant with their first child. It was a difficult pregnancy, and Sue suffered frequent nausea. A doctor prescribed Tigan to combat the waves of sickness that continually washed over her. At the end of 1968 a baby was born. They named him Brian Eugene, after Eugene McCarthy, who was running for president. Brian's birth was difficult, for he was a "blue baby" and had the umbilical cord wrapped around his neck. But he seemed fine. He was a fast learner, and in the summer of 1969, with the television full of the lunar landing, he learned his first word: moon.

A little over a year later, Siegfried was offered the job at Northrop, and Sue continued her doctoral work at the Brain Research Institute at UCLA. They packed up two cars and moved to California. They bought a small, attractive home tucked into a residential neighborhood in the verdant Santa Monica Mountains, just a few minutes from the bustle of Encino, which they planted with macadamia and avocado trees. Gray-green agave plants grew around the house. Southern California was an exotic place compared to upstate New York, a fine place, it seemed, for a burgeoning family.

Almost from the beginning, nothing went according to plan. After Brian was out of diapers and old enough to walk and talk, problems began to show up, something more than the terrible twos. Though he was happy most of the time, he occasionally seemed angry, unsettled, and irritable. Other problems cropped up as he got older. He was a slow learner in school. "When the others are on jump, Brian is still on hop," the kindergarten teacher told the Othmers.

During Brian's fifth year the Othmers had another child, a girl they named Karen. At six months something went wrong: The baby could not keep food down, and she was taken to the Othmers' pediatrician, Dr. Robert Marshall. He recommended tests at UCLA Medical Center at Westwood, a teaching hospital. A number of tests were run, all of them inconclusive. Finally, an exploratory operation opened her skull and found the dark mass of an inoperable brain tumor. Sue and Siegfried were beside themselves. Maybe it was the drugs that Sue had taken for severe nausea during pregnancy—Tigan with Brian and Bendectin with Karen. Doctors hadn't mentioned any possible side effects, but they were left to wonder.

The infant Karen took up residence at UCLA for a battery of treatments, including daily cobalt 60 radiation treatments. Because the radiation had to be delivered precisely to a small part of the brain, Karen had to be kept very still, which was difficult to do with so small a child, and so she was anethesized with chloral hydrate. But the drug made her mildly ill, and that was noted in her chart so an alternative could be used. Yet the next day the technician went for the chloral hydrate, and, again, Karen reacted badly. The following day an intern went to prepare the chloral hydrate, and Sue could not believe it and finally intervened. The intern checked the chart and mumbled, "Oh yeah." "We were learning what it takes to survive in the factory model of health care," says Sue. "As an unsuspecting person, you go into the hospital, you think everybody knows what they're doing. But they don't. And they don't have your interest at heart. I'd spent so much time with Karen that I had met this culture of people who hang around hospitals because they have family members there. And they taught me things, a kind of assertiveness training. I learned to start saying things like 'Where are you taking that child?' 'Who ordered that test?' 'No, I'm not leaving the room.'"

The radiation seemed to help. Karen began crawling, making up for lost time. Then things took a turn for the worse. She started chemotherapy, but it did not help. It was too late, and her voice and awareness of her surroundings faded. Finally, in February of 1975, Siegfried choked out the words "Please do not continue to treat," and Karen died later that same day.

Brian was now six, and his behavior was deteriorating. He had begun flying into rages at school and at home. His belligerence had prompted several kids at school to gang up on him and stuff him into a trash can. Siegfried, raised by a strict Prussian, warned his young son of dire consequences if he didn't behave. Brian frightened him when he snapped: "I'm just an evil person. I'm going to jail when I grow up." Other bizarre comments came out of his mouth: "I'm going to kill myself" and "I'm a warlock." There was nothing in Dr. Spock about children like Brian. "He had changed," says Sue, "from an odd child but a happy one, to one who was confused and depressed and violent." The Othmers nicknamed him Chief Thundercloud for his dark moods. "It was hard," Sue says. "I loved this kid and wanted the best for him, but much of the time I didn't really like him." They thought perhaps Brian was reacting to the death of his sister, or to the absence of his parents for so long during her hospitalization. At one point he had said, "I wish I was in the hospital, so you'd take care of me."

Not that there weren't some good times as well. Sue and Siegfried are avid outdoor enthusiasts, and they spent a great deal of time cross-country skiing and hiking in the Sierras, with Brian in a backpack. A lifelong naturalist, Sue became the unpaid director of a group of docents, or volunteers, who took people on interpretative hikes in the backcountry of the Santa Monica Mountains. The docents also sponsored children's outings, but those came to an end because of Brian's violent outbursts. During one outing, Siegfried looked up to see Brian beating the living daylights out of a little girl dressed in her Sunday whites. Siegfried pulled him off, his arms still flailing, and gave him a spanking. A few weeks later, as Brian was sitting in a shoe store waiting to have his feet measured, he turned to a boy next to him and began pummeling him, until a startled sales clerk intervened. Not long after, the school called and asked for a conference. The principal, Brian's teacher, and other officials unloaded a litany of complaints about Brian. He fought all the time. He had scratched a girl's face, had sworn at the teachers. The Othmers' reaction was visceral. Why was their son, the offspring of two well-educated, liberal intellectuals such a monster? They noticed other parents pointing at them and whispering. Was it their fault? Had they somehow failed their son?

The school asked that Brian leave, and labeled him "emotionally disturbed." Brian was examined by the pediatrician, Dr. Marshall, who suggested an EEG. It confirmed that he had epilepsy. In some ways the news came as a relief. It explained the rages, the doctor said, and the violent episodes. The Othmers began to connect the dots in other ways. Brian had been deprived of oxygen because the umbilical cord had been wrapped around his neck. Perhaps he had suffered brain damage. He spoke sometimes of seeing auras and intense false colors around objects. Once he had gotten into a tantrum at a Little League baseball game and fallen on the ground and made strange noises. Night terrors seized him in the wee hours, and he would jerk violently and cry and sweat and make strange barking noises. The Othmers could not wake him up, so they would simply sit and hold him until it ended.

The diagnosis was a revelation to Brian as well. "He was so elated to have a reason for his behavior," says Sue. "He ran around telling everyone, 'I have epilepsy,' and now he had an explanation of why he was doing these things." In retrospect Sue believes that Brian's bizarre behavior may have been caused by subclinical seizure activity, kindled by the very same source that gave rise to his other seizures, but having effects throughout the brain. "The amygdala governs fear response and is very susceptible to seizure activity," she says. "Seizures here cause terror beyond panic attacks—terror, paranoia, and violent reactions unrelated to what's going on in the outside world."

The doctor prescribed phenobarbital, a seizure medication, for him. It did not help much, and the drug made him dangerously hyperactive. Early one morning, the Othmers awoke to find Brian out on the roof in his pajamas lighting matches. On another day, he disappeared from the house and called two hours later from a friend's house several miles away.

The phenobarbital was discontinued, and Dilantin was the next drug Dr. Marshall tried. It controlled the seizures that Brian had begun having, but the fighting and tantrums and other daytime behaviors continued unabated. The doctor then added another drug for Brian, a new drug on the market called Tegretol. Now Brian's behavior improved quite a bit, with no more violent and out-of-control behaviors, though the drugs left him logy and spaced out. "It helped," Sue said. "But it didn't make him better. It put him in a straitjacket. He could live at home, he wasn't institutionalized, but he was miserable."

Brian, in the meantime, was getting bigger and more physical. One weekend Siegfried took him to Tuolumne Meadows in Yosemite and urged him to climb the steep portion of a large granite outcropping called Lembert Dome with him. Brian did, cautiously and reluctantly at first. When he arrived at the top, he pumped his fists in the air and became a triumphant and confirmed climber. He had discovered the joys of defying gravity and took to climbing everything—trees, jungle gyms, anything. His behavior, however, was still very worrisome. One summer evening at a wine and cheese party at a meeting of the Institute of Electrical and Electronic Engineers, of which Siegfried was a member, someone pointed high up in a redwood tree—a tiny figure was making an ascent. It was Brian, of course, and the Othmers held their breath until he had returned to earth.

Not long after that incident, in November of 1975, the Othmers had their third and last child, Kurt. The Othmers were relieved that Kurt was a healthy baby, though as he grew older, he proved to be a handful.

Meanwhile, Brian had continued to have seizures, and their onset deeply concerned the Othmers. They went to see Dr. John Menkes, a renowned pediatric neurologist and a specialist in these types of childhood disorders. He did an EEG, and Sue and Siegfried went to his office to discuss Brian and the results of the test. They told him about Brian's nighttime seizures and violent episodes and the positive response from Tegretol and Dilantin. The Othmers were shocked at Menkes's response. "He doesn't have epilepsy," Menkes said and blamed the behavior disorder on Karen's death. "He's upset," Menkes said. What about the behaviors and the seizures, the Othmers asked. "How do you know they're seizures?" Menkes said, somewhat irritated and dismissive. The Tegretol dosage was too low to be therapeutic, he said; it was probably changing Brian's behavior as a placebo. Brian sounded merely like a rambunctious kid.

Later, Robert Marshall sent Menkes more information, and eventually the prominent neurologist was swayed; he wrote an apologetic letter to Dr. Marshall, saying, in effect, that Brian did have epilepsy. However, he never communicated his error to the Othmers. "It was a major disillusionment," says Siegfried. "This guy was supposed to be the expert. Yet our observations and experiences counted for nothing." The Othmers had learned another lesson about the factory model of health care.

Over the next few years Brian seemed to tolerate the drugs, and it made life bearable. In 1980, as Kurt was ready to enter kindergarten, the Othmers faced yet another problem. Teachers at the preschool Kurt was attending had urged the Othmers to put their youngest, who they felt was hyperactive, on Ritalin, which is widely prescribed for ADD and ADHD. But the Othmers opposed drugging a passionate if overly energetic four-year-old and found a Waldorf school, Highland Hall, that was philosophically opposed to drugs and believed instead that the answer lay in individualized attention. Both Brian and Kurt were accepted. Brian's lessons were tailored to his strengths, says Siegfried, and he succeeded. "Brian was out of his mind with happiness," Siegfried says. "The school allowed him to climb the eucalyptus tree, and they didn't treat him like he was stupid."

As he entered middle school, Brian proved himself in his academics. But he had a great deal of trouble dealing with ambiguity, subtlety, and humor and in interpreting facial expressions. It was not the way his brain worked. He did not have many friends and was awkward in social situations. He did well in the concrete subjects like science and math but had trouble with subjective subjects such as English. He had fallen in love with nature, and he found peace hiking the mountains and climbing rocks, away from the uncertainty of human relationships.

Through middle school and into high school, Brian seemed to make his way in the world with a little less travail. He was a thin, handsome, and intense young man with a thick head of dark hair. His problems were still serious. Small things—a dropped book, someone in a favorite chair, an intemperate word—would trigger violent, outsized rages and send him off to kick holes into hollow-core doors around the house or smash dishes and other things. Now that he was almost an adult, the violence was even more frightening. There was a great deal of enmity between him and his younger brother, far more severe than the usual sibling rivalry. While the medication helped, it was the lesser of two demons and measurably slowing Brian and fogging his mind. When Brian reached eleventh grade, he began thinking about college. This brought new concerns to the Othmers, who felt he was in no way ready for life on his own. "He couldn't live in the outside world without help," says Siegfried. "If he forgot to take his medication, or was set off by an untoward event, he could have been sent into a self-destructive spiral that he couldn't recover from on his own." The Othmers felt Brian might need a permanent caretaker or would need to be in some sort of situation where he would have supervision.

In the winter of 1985, one of Sue's classmates at Cornell, Susan Rosen, had just returned from the annual meeting of the Learning Disabilities Association in San Francisco, where she now lived. Rosen was struggling with a severely disabled child and searching for help. She had picked up a tape of a lecture by Margaret Ayers on treating epilepsy with something called EEG biofeedback.

Had Brian been one of the kids who take Dilantin for epilepsy and find life dramatically changed for the better, with few side effects, the Othmers would probably have never found neurofeedback. But Brian's problems were still formidable. And because the Othmers had been disillusioned by the politics and flaws of traditional medicine, they had lost much of their respect for the doctor's white coat. Given her background in neurobiology, the idea of changing brain waves through operant conditioning made sense to Sue. Years earlier she had implanted electrodes in the brains of cats at Cornell, to study the EEG correlates of learning. For a while she had continued her work at UCLA, but the pressure of raising Brian and the accidental death of her adviser at Cornell had kept her from finishing. Sue called Margaret Ayers to chat with her about the method. Impressed with what she heard, she ended the conversation by making an appointment. On March 5 she drove Brian to Ayers's Beverly Hills office.

An EEG is somewhat similar to a computer that diagnoses automobiles. When a mechanic fixes a car, he often hooks the vehicle up to a computer, and the computer screen shows where in the system the problems lie. In the case of EEG biofeedback, the EEG readout shows where there are abnormal brain waves. This was the era before computers had become part of every aspect of life. Ayers hooked Brian up to an old-fashioned, analog, pen-and-ink EEG machine. A normal brain that is awake and performing tasks is desynchronized. That is, there are different electrical signals firing in many different places—senses, memory, movement of different parts of the body—to create the symphony of everyday life. In an epileptic seizure the brain ceases to do all of these myriad functions. A large, uniform wave of low-frequency activity, like a giant ocean swell in a storm, washes over the brain, and all the neurons are recruited to fire at low frequencies, to the point of disrupting consciousness. Brian's EEG was markedly abnormal, with excessive slow wave activity. The training started on the left side to teach Brian to strengthen his brain by holding those slow waves at bay.

Brian was hooked up to the little black box that Sterman had used, with two lights, red and green. To begin with, Ayers placed the electrodes on the C3 site—between the top of the ear and the center of the head on the left side. Brian shifted his state from paying attention to relaxing his focus and deep breathing, until he made the green light come on and the red light stay off. Ideally, the thinking mind is taken out of the process, and reflexes take over. After practicing, you do it without thinking about it.

Neurofeedback was something Brian wanted to do because he did it well. A desire to make changes, and to get well, is a critical part of the process, experts say. Engaging the will somehow helps a person make changes in the EEG through the thalamic generator. Brian did two sessions a week, for thirty minutes each time. Approaching the treatment scientifically, Sue began keeping a diary, a small red, leather-bound book with gilt-edge pages, filled with brief day-to-day observations of Brian's biofeedback training and progress.

After just a few sessions, Brian seemed different, better somehow. He was smiling more and seemed to be thinking more clearly. Yet the Othmers held their breath: they were concerned that they might be imagining things, because this, of course, was what they wanted to see. Eight sessions into the treatment, they had their first independent verification that something was happening. And it was a blind test, for they would be spending the weekend with friends, and no one there knew Brian was training his brain. At the end of March, the Othmer family went camping on a docents' weekend at Point Mugu, a beautiful piece of wild country overlooking the Pacific Ocean. A brief entry in Sue's diary tells what happened: "Brian cut his finger badly. He was acutely embarrassed but under control. Friends noticed that he was 'lighter.' We had noticed by this time that the irrational outbursts had stopped." The observation by friends validated subtle changes the Othmers were noticing. Brian seemed to have more energy and did not seem to fight quite as much with his brother.

The sessions continued. A few sessions later, in early April, the Othmers took a family trip to Monterey to see the new aquarium. Brian did not want to go. At first Siegfried agreed, having experienced long, difficult car trips with two children, one with ADHD and the other with a personality disorder, fighting like a cobra and a mongoose in the backseat. Even though Brian was seventeen, the Othmers were also uncomfortable about leaving him alone at home. Sue and Siegfried insisted he accompany the family. He was surly and kept to himself, but only for a short while. "He was repeatedly drawn into conversation, despite himself," recalled Siegfried in some notes that he jotted down at the time. "Another time we looked into the backseat, and they were asleep on each other's shoulders. It had never happened before." The Othmers were finally allowing themselves to become hopeful.

The sessions continued. Some notes from Sue's journals:

> April 7, on the trip to Monterey. Brian's good spirits continue to make this the best trip ever.
>
> April 12. Brian is coming out of his room in the evening to engage someone in conversation. He is hanging around after dinner to talk and socialize. He is smiling and laughing and can even take a joke.
>
> April 17. Brian is acknowledging changes in himself and believes he is concentrating better.
>
> April 20. Brian is planning ahead. He is thinking about a summer job to earn money for his projects.
>
> April 23. Brian and Kurt are walking back from the school bus having conversations these days, instead of Kurt arriving alone in tears.
>
> April 30. Margaret says Brian is doing well; she would like to start reducing his medication.
>
> May 3. Brian expects to finish his main lesson book—the first time ever. He expressed an interest in going to parties, the first time since 8th grade.
>
> May 28. Recorded EEG. Observed considerable reduction in EEG amplitude and in incidence of low-frequency activity.
>
> May 29. Saw Dr. Marshall. He thinks Brian must be outgrowing his learning disabilities. It remains for him to be convinced, but he is willing to reduce the Dilantin dose (though not the Tegretol).
>
> June 3. Marshall is willing to give [biofeedback] a chance if he can dictate changes in medication.
>
> June 5. Reducing Dilantin from 300 mg a day to 260 per day.

June 9. Brian is much more responsive to Kurt. He doesn't always shut him up with "I don't want to know" or "I don't care." Kurt built his Lego train into a message delivery system in his cardboard fort. He sent Brian a message: "I love you." Brian returned with his own message: "I love you too but I have work to do so bye." We have come a long way from "Brian is a lion. Kirt is dirt."

July 1. Dr. Marshall okayed a decrease of another 30 mg to 200 per day.

September 18. Decreased Dilantin dose to 100 mg a day.

September 29. Brian is incorporating regular exercise and clarinet practice into his schedule.

October 19. Brian has much more energy after reducing Dilantin dose. He is joining school cross-country team.

October 30. Dilantin level reduced to zero! (61 sessions.)

November 5. Brian complimented Margaret on her clothes!

November 8. Brian enthusiastically and on his own initiative bought a watch for Kurt's birthday. Such interest in someone else's desire and happiness is a welcome change.

It did not take this long for the Othmers to be convinced. By the end of the first month of the training, they were flummoxed. Neurotherapy was amazing! It had done things for them in several months that years of top-flight and costly medical care had not come close to doing. It had given their son—their real son—to them. All this time he had been trapped inside a damaged brain, and now he had been shown a way out. Why didn't more people know about it? They took Kurt to see Margaret for his ADHD. After a single session his bedwetting stopped and did not return. Further training calmed him down considerably. Sue began training for her own hypoglycemia. Low blood sugar caused her chronic fatigue and frequent illness, and it was one of the things that had forced her to curtail her Ph.D. studies.

After her first session, she experienced the clean windshield effect, the quiet energetic feeling some people have that lasts for a few hours after a session. When Siegfried came home from work, she told him, "There's a new person in the house." He looked around, mystified. She grinned: "It's me." After a series of sessions her symptoms of low blood sugar had substantially disappeared. Brian was making friends for the first time. The concept of training the brain to be stronger blew the Othmers away. "We knew right away that this was something we wanted to be involved with," says Siegfried. In May, six weeks after Brian began training, Siegfried began talking with Ayers about modernizing her technique with a computer system and forming a company to market the equipment. He talked to a friend of his and Sue's, named Ed Dillingham, who was an expert in writing computer software.

Brian graduated from high school in the summer of 1986. The following fall he started college at Cal Lutheran, a small private school a half hour away. The dullness brought on by the drugs had disappeared from his eyes, and he seemed more alive, more present. For the first time since childhood, he seemed happy. There were still problems, but the Othmers believed that Brian would be able to have some semblance of a normal life.

CHAPTER SIX

EEG Spectrum Takes Flight

∞

On October 29, 1985, as Siegfried recalls, he, Ed Dillingham, and Margaret Ayers arrived at a verbal agreement to form a partnership. Margaret would provide the expertise; Dillingham would write the program; and he and Siegfried would help build and peddle a new and improved computerized version of a neurofeedback instrument. It would take off, everyone was convinced, once the instrument was more powerful and easier to use. Ayers would also be able to get out from under the onerous monthly payments she was making to lease her equipment. Exuberant at the prospects of their new business, the partners sealed the deal with a celebratory dinner at the Cheesecake Factory in Beverly Hills. Their association, though it eventually bogged down in personal and professional conflicts, would take neurotherapy out of a few clinics and introduce it to a much larger audience.

Working nights and weekends, it took more than two and a half years for Dillingham, a white-haired, white-bearded bear of a man who has since died, to write the software. In December of 1987, the first computerized neurofeedback instrument was placed in the bedroom of Dillingham's home for its maiden voyage. It was a single computer that read and displayed the EEG on the screen instead of the old pen-and-ink–based system; there was also a black box with lights, similar to Margaret's unit. A green light got brighter as the beta, the 15 to 18 hertz, was increased. (The Othmers did not know about SMR, the 12 to 15 hertz, at this point; Ayers hadn't told them.) If the user dipped below 15 to 18 hertz, the one red light came on. In addition to Dillingham's time, the unit cost around $15,000 to build.

Meanwhile, the relationship between the Othmers and Ayers began to sour. She claims that she never wanted or agreed to a partnership. The Othmers claim she did and then she realized that if the alliance progressed, she would soon have to share her singular knowledge and autonomy with her partners. The disagreement surfaced in one confrontation, recounted by the Othmers. They had treated Brian on the test instrument, and Siegfried says Ayers was outraged that they were treating without her. And she didn't want Sue on the board, which Siegfried insisted on. Things were starting to unravel. Dealing with Ayers became frustrating, the Othmers say, as she took days to get back to them when they had questions. Part of the problem, as even Siegfried admits, was that the partnership was a lopsided arrangement. Ayers had the lion's share of the intellectual investment, Dillingham had his computer expertise, and Siegfried had what he calls "hot air": little more than industrial-strength enthusiasm to hang spotlights all over the technique to bring it to the attention of the world.

Siegfried recalls a tense meeting in October 1988 that he and Dillingham held in Ayers's lawyer's office. The meeting was with each side's lawyer to establish a corporation, but Margaret didn't want Sue there and said her presence would be a deal breaker. Her demand was unreasonable, Siegfried felt, and Sue went anyway. Tension hung in the air, and lawyers for both sides hastily conferred. Afterward, each side was to meet with its own lawyer. Sue and Siegfried moved toward their lawyer, Ayers toward hers. Dillingham, Siegfried says, did not join either group right away and looked back and forth, not sure which way to move. The Othmers' attorney asked him, "Am I your attorney?" Dillingham said yes and joined the Othmers. "It was at that moment," Siegfried says, "that Margaret realized that she was not simply getting rid of the Othmers, who she no longer needed in the partnership, but was also losing Edward Dillingham, on whom she expected to continue to depend."

Ayers vehemently disagrees with the Othmers' account and became livid when I asked her about her partnership with them. "We did not enter into an agreement, verbal or written," she said. "I treated their son, they said they were grateful, appreciative, and they would help get this work out all over the world. They took my ideas, went off with them." She says she did not consult with them on the construction of the new instrument.

Whatever type of relationship had developed dissolved bitterly, and a poisonous atmosphere among yet another set of primary players was created, and would come back to haunt the Othmers as they struggled to start the business. Despite the parting of the ways, "I am to this day grateful that Margaret was there," says Sue Othmer. Sue also believes that Ayers's claims about the effectiveness of treating comas, head injuries, and other serious problems are very real.

For the Othmers and Dillingham, the breakup with, as far as they knew, the only clinician in the field, was a body blow. Yet they—as well as Ayers—possessed the updated technology, and they were still infused with evangelical fervor, interested in getting this idea to the public and breaking up what they saw as Ayers's intellectual monopoly, which had kept the technique under wraps. "It took us a very short time to realize we weren't going to walk away from this," says Sue. "Knowing what we know, and knowing this is going to happen in the world, we would just die watching someone else do it." After the breakup with Margaret, the three others drove to the Othmers' home and sank into the couch in the living room, dejected. Siegfried and Dillingham turned to Sue, who, despite having a background in neurobiology and a confident bedside manner, had been only peripherally involved. "It's your turn," Siegfried said. The next day EEG Spectrum was born. "By throwing us out, Margaret created the very thing she feared most," says Siegfried, "because by this time the hook was so deep in us we weren't going to let go." Perhaps most important, the Othmers weren't in it just as clinicians. They were bringing a unique evangelical passion with them. "We're parents," says Sue. "We know what people are going through. We know it works. We don't need any more studies—we need to get this out there."

They were fueled by a couple of other things as well. First was their naïveté. All that was needed was for the world to hear the gospel of neurofeedback and it would, they assumed, take off. Letters to the right people and the press, show-and-tell, and workshops would be a match to gasoline. "We were like the bumblebees that are aerodynamically unable to fly but don't know it, so we flew," says Siegfried. Second was Siegfried's irreverent—some say bombastic—nature. He does not abide by the rules of engagement in scientific discourse, such as not hyping your own work or making claims for which there are not several careful studies or not seeking attention from the press. The time for that was over. Neurofeedback had been under wraps far too long because people were afraid to step out and make bold statements. Not Siegfried. He and Sue had both had enough of "medical science" and knew of its ample shortcomings. And Siegfried, as a veteran Ph.D. physicist, was well versed in the idiosyncracies of the scientific mind and the bickering, attacking, politicking, and backbiting that goes on under the rubric of scientific investigation. Many mainstream scientists, he says, are anything but open-minded, and people who are well respected in their field are often deeply threatened by knowledge of their subject that might lie outside their carefully constructed view of things. "Physicists have a general feeling of fearlessness," says Siegfried. "They feel entitled to speculate on any topic. It's irritating to have physicists rush into your field and start pontificating. The fact of the matter is when you understand the laws of nature, you can do that." It is the very fact that he and Sue are outside the fields of medicine and psychology, Siegfried believes, that has made neurofeedback happen, because people in those fields are trained not to believe in such things or even to examine them with an open mind.

Dillingham had the prototype machine at his home, and Margaret had the only other one in existence. EEG Spectrum built a third, and the Othmers set it up in Brian's bedroom (which, now that he was away at college, the Othmers used for guests) and started inviting friends over to try out the new brain wave machine. "We had lots of takers," Sue says. The Topanga Canyon docents group provided a ready source of subjects, and other friends also volunteered their brains. But the breakthrough came when a friend named David Wells, another parent at the Waldorf School and a chiropractor, offered his eight-year-old, extremely hyperactive boy as a guinea pig. "This kid was sweet but really wild," says Sue. "Breaking the legs off furniture and stuff." They sat him down in the guest room, and he began training his brain and trying to kick the computer at the same time. "It was serendipity," says Sue. "Within ten sessions he was miraculously transformed with beta, and that led us on. If he had been an SMR kid and we'd screwed him up, we wouldn't have been as positive."

For a long time there were two basic approaches to neurofeedback. Generally, those who do beta, or 15-to-18-hertz training on the left side of the brain, feel alert, and those who do 12-to-15-hertz SMR training on the right side feel calm. Most hyperactive kids respond better to SMR training; a minority do better with 15-to-18-hertz. The Othmers didn't know about SMR at the time; they trained only 15-to-18, which is what Brian, Kurt, and Sue had received. If the eight-year-old boy had gotten more hyperactive with the beta, the Othmers would have been demoralized. As it was, they were elated and set about spreading the word. "If we were on an airplane," says Siegfried, "the person next to me didn't have a chance. If I was next to their ear, they were going to get it. By the time they got to their destination, they would know all about it."

A few months after they began, the Othmers started treating people for $40 per session. Clients had to make their way through the kitchen and living room to be hooked up in their guest room. Friends and friends of friends came for training. The wife of the Othmers' minister knew some kids at the church who had learning problems, and she referred them. On December 5, 1988, the first EEG Spectrum office opened in San Francisco. It was run by Sue Rosen, the friend who had given Margaret Ayers's tapes to Sue. A week later they opened a second office in Encino, where a chiropractor friend used his office only part-time and offered them a time-share to do the training when he was not there. Sue Othmer and a nurse, Sandra Shapiro, were the clinic's only employees, and they offered one protocol—on the left and right sides at the 15-to-18-hertz range. Sue was also trying to fold SMR into the mix. She did a year's worth of training at the Biofeedback Certification Institute of America in LA, the industry's regulating body. It wasn't a course on neurofeedback, of course, but on other types of biofeedback, and she was now a certified biofeedback practitioner.

One of the first things the Othmers did after splitting with Ayers was call Barry Sterman and his wife, Loraine, and invite them over for dinner to talk about the business. Siegfried says Ayers had warned them away from Sterman. Now that they were on their own, they felt that his name would lend credibility to EEG Spectrum and help make inroads into the biofeedback community. Sterman was skeptical about Spectrum's claims for such a wide application but was interested and would later come to work for EEG Spectrum.

By now Siegfried had moved to Hughes Aircraft, where he was a laboratory scientist working on infrared imaging for antitank missiles. In February 1989 he quit that job to work full-time as president of EEG Spectrum. Siegfried's main task was to notify the powers that be of this wondrous new technology, and so he sat down at his computer and wrote letters, dozens of letters, to leading researchers, especially in the field of ADD and ADHD. "I thought the world was waiting for this," he says. As time passed, however, there was no response to his letters. "None," he says. "Zero." He could not figure out why people weren't even curious enough to come and see it or to hear more. Now he realizes that people thought he was a crank. He asked his doctor, whose son had dramatically improved academically after training, to introduce him to the psychiatrist to whom he referred patients, so Siegfried could talk to him about the possibility of using the technique. When the psychiatrist found out who he was, he canceled the appointment. Such setbacks provided lessons on just how difficult introducing this new technology to the world was going to be.

The struggle has gotten easier, though it is by no means over. While Sue battles in the clinic, hooking people up and advancing the clinical knowledge of neurofeedback, one of Siegfried's roles, as president of the company, now called EEG Info, is the big picture. He functions as a kind of philosopher general, head of a ragtag band of rebel forces arrayed against mainstream medicine, lobbing metaphors and analogies about the failure of mainstream science to be open-minded and its propensity to instead attack those who have different ideas. His years of bucking the system and supreme confidence in his judgment have given him plenty of fodder for comments on the nature of scientific politics and the mindset of scientists. Siegfried believes deeply that neurofeedback is a revolutionary system, a powerful and precise way to manage the human central nervous system, something that is currently managed in only a gross and haphazard way, primarily with drugs. It hands responsibility for that management to the people themselves. It democratizes one of the most important things on the planet: the human nervous system. The change is so dramatic, he says, that it is akin to what biologists and paleontologists call punctuated equilibrium, a sudden and powerful burst of evolution off in some isolated part of the population that gives it an advantage, while the rest of the population continues on its plodding course. Such leaps, he says, don't come from within the field; they come from outside or on the fringes. And they are not welcome, for scientists have a great deal invested, financially and psychologically, in the body of knowledge as it exists. That is why the scientific community first ignored neurofeedback, and later a few members of the community virulently attacked it. He expects more attacks. "It's an immune reaction," he says. "The antibodies want to keep the contagion from spreading into the body of science, and so they swarm all over it and try to kill it."

The Othmers had no idea how tortuous a path lay before them. Even their critics admit that they are extremely bright, well-spoken people. And though their claims are, on the face, incredible, their arguments are usually logical, cogent, carefully articulated, and well supported by research in other areas of neuroscience. They have a coherent unifying model, and the results they get are predictable. "Forget what you have heard about biofeedback and take a serious look," says Siegfried. "That's all we asked." It is thus remarkable how few people they could find to take them up on that, the antipathy they engendered, and how well fortified the wall around mainstream science is to repel anything considered fringe. They had trouble finding a friendly ear anywhere in the field, even including Sue's older sister and her husband, a Harvard trained psychiatrist. Her sister has a Ph.D. in neurobiology from Harvard, had done pioneering work on the blood brain barrier and immunology, and had to battle the medical establishment to get her views across. Ironically, hers was the same kind of fight the Othmers were involved with, and yet, "They couldn't talk to us about it," says Sue with a touch of sadness. "They saw Brian, and he was utterly transformed. It was so hard to miss. He was a different person. We didn't see them a lot, just occasionally, but they saw it. And it was just so embarrassing to have these California kooks in the family, they couldn't even talk to us about it." "We were slaying all these sacred cows," says Siegfried, "and we didn't even know they were sacred cows."

The first official "unveiling" of EEG Spectrum's machine to the biofeedback community was at the Town and Country Hotel in San

Diego in 1989 at the annual meeting of the Association of Applied Psychophysiology and Biofeedback (AAPB), the establishment organization of biofeedback. At the time, AAPB was concerned almost exclusively with relaxation training: hand warming, galvanic skin response, and muscle work. The Othmers expected to light the place on fire with their revolutionary new equipment and to start selling the $19,000 instruments; instead, they were barely noticed. Biofeedback professionals looked at the instrument, nodded, and shrugged. People had heard of Sterman's work, but few were interested enough or knew enough to use the approach. There was good reason the field wasn't interested. Biofeedback practitioners were laboring mightily to birth their field into the mainstream and were still living down the reputation alpha training had earned in the 1960s and 1970s. Here were two outsiders no one in the field had ever heard of raising the specter of brain wave training again. Traditionalists were, and many remain, understandably concerned that claims of a cure for epilepsy and ADD could deal a serious blow to the field.

But that was not the most difficult part of the weekend. As the Othmers sat over scrambled eggs and toast one morning in the hotel dining room, a friend of Ayers walked over and handed them a copy of a lawsuit that Margaret had filed against them in state court for theft of her machine. They were due in court the following Monday. Shaken, the Othmers went to the booth where they were demonstrating their equipment. Someone came over and said, "Oh, you're the ones with the lawsuit against you." A copy of the lawsuit had been hung on the bulletin board. No one would do business with them. Siegfried called a lawyer to see what he should do. "Take it down," the lawyer said. Siegfried did. "It was cannon fire just as we're popping up out of the hole," Siegfried says. "At this point Margaret still hopes to contain it."

Meanwhile, with the world not seizing neurofeedback the way they had hoped, the Othmers were looking toward other strategies. Treating people in the Spectrum clinic would obviously help people but would do little to bring this success to the world's attention and spark a revolution. Absent the ability and the funding to do double-blind controlled studies, they also needed a way to validate their claims.

To accomplish that, the Othmers decided to create a network of people trained to use the equipment. "We needed to have mainstream professionals replicating our work," says Siegfried. "If enough people, trained people, are getting the same results, then double-blind studies don't matter. We planned to build demand from the ground up, instead of working from the top down." An affiliated network of professionals would also push the envelope on what the technique could accomplish, as hundreds of practitioners tied together by the Internet tried myriad approaches for different problems and shared clinical results and other information. And it would create a market for their instruments.

In August of 1990 the Othmers launched their professional training course to teach the therapy to psychologists, MDs, therapists, psychiatrists, nurses—basically, any professional who was interested in learning. It was an opportune time, for a larger patient-led health care revolution was under way, as acupuncture, herbs, nutrition, chiropractic, and a host of other alternative therapies made their way into the mainstream.

One of the first people to attend a training session was Steven Rothman, a Ph.D. psychologist from Bellevue,Washington. He had practiced muscle biofeedback for more than twenty-five years and was interested in brain wave training, but the word among traditional biofeedback practitioners was that neurofeedback did not work that well. Then he attended a lecture by Joel Lubar and was impressed with the claims for helping people with ADD. But it was not until he took the Othmers' training course, says Rothman, that he learned that neurofeedback would overhaul the way he practiced psychology. "I was very skeptical," he says. "If this was such good stuff, why wasn't anybody else doing it? I told my wife, 'If we don't see the same results, we have a very expensive boat anchor.' And we don't have a boat." In the last decade, neurofeedback has become 90 percent of his practice. He treats ADD, fetal alcohol syndrome, depression, autism, and mild head injuries, and estimates at least 70 percent of his patients make substantial gains. A large part of Rothman's business is home trainers. He leases machines to people who live a long way from the office for $325 a month and $125 for a required hour of consulting. Through faxes and e-mail he can keep track of their training, and the machines are programmed to stop working after the appropriate number of sessions. "It's like a prescription for medication," he says. "It runs out, and they have to call me to get it refilled."

In April of 1990 Margaret Ayers filed a patent infringement lawsuit against the Othmers in federal court, adding it to the state lawsuit. Unbeknownst to them, she had patented the equipment. She also asked the judge to issue an injunction, which would mean that they could no longer sell the equipment. But the judge refused that request. Ayers now had a problem: if she waited for the lawsuit to come to trial, the Othmers could keep selling the equipment in the meantime. So she agreed to binding arbitration, which she and the Othmers carried out in the summer of 1991. Ayers lost on a key point. The arbitration panel ruled that Siegfried Othmer and Ed Dillingham had an equal ownership in the software with Ayers. But it was a costly victory for the Othmers—already struggling to survive, they were forced into a second mortgage on their home and had to borrow an additional $65,000. All told, their legal fees were around $200,000.

A few positive things happened along the way to keep them going. Throughout the course of their efforts to build their business, sporadic media attention helped them stay afloat and even led to new developments. The first episode of that kind happened in the spring of 1991 when the magazine *Woman's Day* did a story on Joel Lubar's work with ADD and ADHD children and reported that neurofeedback had not only decreased impulsivity and distractibility, but even raised the IQ of these children ten to fifteen points. The story caused a flurry of calls and appointments for training.

Siegfried also did a study of his own, a clinical outcome study, which means that they measured results before and after testing, though without a control group. Working with a psychologist named Clifford Marks, they found eighteen hyperactive children. They took careful measurements before training, and after forty sessions of neurofeedback they measured again and found across-the-board increases in such things as self-esteem, legibility of handwriting, organization skills, speech and verbal expression, and an average increase of twenty-three points in IQ. The results were not published. In fact, the Othmers had not published any of their data in a refereed journal until 2005, and that has fueled their critics.

The Othmers were also getting more firsthand experience with the training. In 1987 Sue's father, Joseph FitzGerald, was diagnosed with progressive super nuclear palsy, a degenerative brain disorder with symptoms similar to Parkinson's disease, including loss of memory. Neurotherapy does not treat the structure of the brain; it treats functional disorders. If the brain tissue is severely damaged, there is nothing neurofeedback can do to reverse it. The best effects are seen in those cases where the brain is physiologically sound but not, for some reason, operating as well as it should. But neurofeedback, practitioners say, can improve function, even in damaged brains, and buy a better quality of life for a longer period of time. Sue's father trained first with Margaret, before the breakup, and later with his daughter, throughout the long, slow decline of his illness. "It was startling," says Sue. "He was having significant problems. He didn't want to speak because he would get lost in what he was saying, so he was kind of depressed and withdrawn. On the way back from the first session, he started to point things out and engage with us, to wake up. The training had enormous impact on his alertness and depression." Eventually, as the disease progressed, he had his own instrument and trained every day. If he was away from the training for a week, his wife, Ruth, would notice his functions continue to slip until he could get to the training again. As his brain deteriorated, he continued to train, and Sue estimates that it prolonged his cognitive abilities for two or three years beyond what normally would have been the case.

Against this backdrop, meanwhile, Brian Othmer continued to establish his own life at college. After two years at California Lutheran University, he took a semester off in the fall of 1988 to climb and backpack in the Rocky Mountains of Utah and Colorado with the National Outdoor Leadership School. Then he transferred to California Polytechnic Institute in San Luis Obispo, where he declared a computer science major. He continued to improve with his training, which he was still doing once a week. Not everyone is sensitive to biofeedback, but Brian was extremely so and could describe very well and in great detail the effect of training on his brain. "He would come into the office, and I'd say, 'How do you want to train today?'" says Sue, "and he'd say, 'I have too much frontal delta.' He knew what was going on in his head." Brian's experience was immensely important to the Othmers and their early efforts to understand the nature of neurofeedback. Brian was also captivated by the technique and how it had transformed him. While studying computer science in college, he designed new neurofeedback programs for Pac-Man and Boxlights, two games EEG Spectrum still uses. Brian's experience also revealed to the Othmers the limitations of EEG biofeedback. While they believe it is the single most powerful tool for many problems, alone it was not enough to treat their son; the Othmers needed to integrate their approach, especially with problems as severe as Brian's. Drugs, for example, remained part of his treatment package. Although he had stopped taking Dilantin, he continued to take Tegretol, albeit at a much lower dose than he would have otherwise. He needed to carefully manage his sleep. He couldn't stay up late if he needed to get up in the morning for school, for it weakened his body and made him more susceptible to seizures. There were diet-related problems as well. He had allergies to allspice, paprika, and chocolate, which stimulate the nervous system. He had experimented with peppermint schnapps in his dorm room and found that alcohol also weakened him and made him more susceptible to seizures. Though he still had problems—his illness was a very serious one—he had truly been transformed by neurofeedback.

On March 1, 1991, Sue called Brian and left a message on the answering machine in his dorm room. He didn't return the call, but that was not unusual, for Brian had fully embarked on his own life, and they often did not hear from him for weeks at a time. Then on March 6, two uniformed police knocked on the door of the Othmer home. "There's been a terrible accident," one said. "They found Brian dead in his room at CalPoly." Siegfried invited them inside. Brian's roommate had noticed that Brian was still in bed when he left for classes, they told the Othmers. When he returned that afternoon, Brian was still in bed. He shook him: no response. His roommates had heard him have a seizure the night before, about one A.M. An occasional seizure was not unusual, so they thought nothing of it. But this one had either stopped his heart or his breathing. Sue believed that Brian so hated the effects Tegretol had on his body and mind that he was constantly trying to shave down the dosage, beyond, apparently, a safe level. His journal is full of his efforts to pare back his dosage to escape the side effects of the drug. "It seemed real and not real at the same time," says Sue. "Brian was such a tenuous presence anyway. It was such a miracle that he lived as long as he did. You had that sense of 'Of course, it could have happened so many times before.' Yet at the same time I couldn't take it in. I couldn't engage with it."

Ayers has raised the notion, in court and at other times, that the reason Brian died may have been the fact that the Othmers have no medical background. And she has denounced their lack of training to me. "I healed their son," she said. "And they undid it with their arrogance. And because of that he's dead." But there is no medical certainty as to why Brian had a seizure and died. "It's true that we know a lot more now, and we would have trained differently," says Sue. "But the reason Brian died is that he chose not to take medication at the level he needed. He chose to have seizures. He was absolutely determined to get off the medication. And he felt that part of training himself not to have seizures was to walk that edge and allow himself to have seizures. He was imprudent."

The Othmers drove to San Luis Obispo and returned home with Brian's ashes. They sat down and read through a journal they found among Brian's belongings, a detailed, introspective record of his experiences as he moved away from home. A memorial service was held for him at the Bel Air Presbyterian Church. Friends of Brian read from his journal while his music teacher played on Brian's Native American flute. Several friends read poems they had written in his memory.

Two months after Brian's death, Sue's father died from his brain disorder. He was so ill he had not been told of Brian's death.

A heavy twilight gray descended on the Othmers, and it was one of the few times the couple considered throwing in the towel. "It really put us into a tailspin," says Sue. "There's this leaden depression that comes from this, and for a while I could barely function. We were doing the training, and thank God we were in the stress reduction business, but, nevertheless, it was still extremely difficult." Yet to abandon the struggling field now, they felt, would be the worst kind of betrayal of what they had come to believe. Brian's life had become a parable to the Othmers, something they could hold out to other people to teach the simple yet powerful lesson that children like him are not by nature lazy, violent, excessively angry, somnolent. They are shackled to imperfect brains, which in itself is a revelation, but, moreover, they have brains that in many, if not most, cases can be tuned to work better. If this is true—and the Othmers have no doubt—it is a revolution. So much of the world's suffering, they feel deeply, is unnecessary and reversible. They remember Brian's elation as a little boy when he learned he had epilepsy and ran around telling everyone, delighted that he had a reason for his behavior. He wasn't bad or evil or a warlock. There is not a more essential lesson. To give up, the Othmers felt, would have been to abandon much of the work Brian had done. Before he died, they asked him about the possibility of telling his story, and he said he wanted it told, "warts and all," as he put it. "Brian's story is the story of our future, as individuals and as a society," says Siegfried. "The twenty-first century will be about biology. Our own biology. We own our central nervous system to a much greater degree than we ever thought possible. That is the story that Brian has to tell." Devastated emotionally by Brian's death, these thoughts gave the Othmers the will to keep going, just as an inheritance from Sue's father provided the financial capability. "It was a godsend to us, and we lived on it for a year," says Siegfried. "It was just in time."

Things moved slowly for much of 1992. Then on January 12, 1993, the gray clouds parted. The ray of sunshine came in the form of the *Home Show*, a now defunct national morning show on ABC, based in Los Angeles. The show's producers wanted the Othmers to talk about and demonstrate their work. There was a problem, though. Sue had contracted cancer of the colon and was still recovering from surgery. Siegfried was tapped to be on the show. The television producers built a room with a one-way mirror to film a nine-year-old girl who was receiving the training with a hidden camera. Everything was set up and ready to roll. Then, just before the segment was about to air, the power went off and plunged the studio into darkness. "I went into orbit," Siegfried says. "The power came on with seconds to go to airtime, and I'm supposed to be calm and collected and know what I'm saying."

But the segment couldn't have gone better for EEG Spectrum. It opened with a teenage girl talking about her battle with ADHD and her struggles in school. Thirty sessions later, she said, her learning difficulties had vanished. Then the younger girl was filmed as she played Pac-Man to treat her ADHD. The hosts and the show's consulting pediatrician, Dr. Jay Gordon, talked with Siegfried from the couch in the living room–like set. After they showed the little girl doing the training, they asked Dr. Gordon what he thought. "I'm trying hard to be skeptical," the doctor gushed, "but it looks very, very good. It's very close to mainstream medicine. I wish they would publish more studies in recognized journals so I wouldn't have to argue with my colleagues." One of the hosts asked the doctor if he felt neurofeedback should be added to the arsenal of treatments such as drugs and therapy. No, the doctor said, "I would like to see biofeedback be used first." And throughout the show the hosts referred to Siegfried as Dr. Othmer and never once mentioned that he was a doctor of physics, not a physician or a Ph.D. in a germane field.

The fledgling EEG Spectrum soon received over 5,000 letters. "We had a hundred and forty-three letters appear the very next day, some from as far away as Indiana, and those people would have had to run to their mailbox, get it right in, in order for us to get it that day," says Siegfried. "We had people drive up to our door. They didn't have a phone number, but they had an address." The exposure caused interest from professionals to soar. But in some ways the boost occurred too soon. The clinical training aspect of the business was new, and there were perhaps a hundred affiliates, which is not many when they are scattered across fifty states. But the blizzard of interest provided a much needed boost in cash flow.

A month after the *Home Show* was the first Winter Brain Meeting, in Key West, Florida, a meeting organized by Rob Kall, a biofeedback practitioner and head of a small company called FutureHealth, based in Newton, a suburb of Philadelphia. While the AAPB meeting is a large gathering of all kinds of biofeedback practitioners, the Winter Brain Meeting is for the cognoscenti of brain wave training as well as other new, experimental health care techniques. The leading figures in the field and related fields—Les Fehmi, Len Ochs, Joel Lubar, Joe Kamiya—were all there. The attendees include many of the same people who pioneered alpha training in the 1960s and 1970s. Kall is credited with creating what turned out to be a seminal meeting, a jolt of adrenaline into the tiny field. People had been working separately for years, and this was the first time they had all gotten together to share ideas and research and to get a sense of the big picture of brain wave training. There was excitement about the fact that the field had received some media attention, via the *Home Show*.

It was a stimulating conference. There was a feeling that the field might be coming into its own. The most controversial part of the gathering was the Othmers' presentation. Even though many of the people at the conference were doing EEG biofeedback, and a few like Joel Lubar were doing high-range training for ADD, no one was making the wide variety of claims that Siegfried Othmer was making in his inimitable and uncompromising way. Amid the tropical Florida breezes at the Holiday Inn, Siegfried boldly told the group that neurofeedback could be used to treat everything from epilepsy to premenstrual syndrome, to closed-head injury, to depression, to strokes, to Tourette's, and on and on. He said they had treated children as young as two. Though many people were intrigued by what he had to say, not everyone was. As Siegfried continued, it was too much for Joel Lubar, who stood up and, angry about the lack of research to substantiate the claims for efficacy beyond ADD and epilepsy, shouted Siegfried down. As the two engaged in hot debate, George Von Hilsheimer, an EEG clinician in Florida, stood up and loudly said, "I came to hear what this man has to say." Siegfried finished. Now there are more people doing high-range brain training than any other type of biofeedback. (The Winter Brain Meeting ended in 2008. That's too bad. It had attracted intelligent, out-of-the-box thinkers who are a hallmark of the field and who have done a wide variety of research. The seminars were fascinating, and the talk at the bar was the most unusual of any conference I attended.)

When the Othmers returned to Los Angeles flush with the attention they received at the Winter Brain Meeting and the *Home Show*, they felt they had goosed along their brand of neurofeedback substantially within the ranks. But the trials of EEG Spectrum were far from over. They had begun in May of 1992, like something from a biblical nightmare, when the first of a series of disasters swept Los Angeles.

Soon after a jury returned a not-guilty verdict in the beating of Rodney King, rioting, looting, and arson tore Los Angeles apart. Fifty-two people were killed. Then, in 1993, brush fires raged across 200,000 acres of wildland and consumed more than eight hundred homes. Ensuing torrential rains caused massive flooding. Siegfried and one of his instruments were almost swept away when he arrived at an intersection just minutes after flash floods had swept several other cars into the Sepulveda Basin. In 1994 the Northridge earthquake struck the San Fernando Valley and caused $30 billion in damage. The disasters created huge blocks of time when people did not venture out much, and the incidents deeply chilled the clinical income. And all of this happened against the backdrop of a deep recession in the aerospace industry. The Othmers did what they had done in the past: lashed themselves to the mast and sailed through the storms, continuing to rack up credit card debt.

Perhaps one of the cruelest blows of all occurred in the mid 1990s. Siegfried had met with a child psychiatrist at UCLA named Dennis Cantwell, one of the leading experts in ADD and ADHD, and had convinced him to take a serious look at neurofeedback. After several meetings Cantwell agreed to design and participate in a three-year, $3 million study of 150 children with ADD and ADHD. Cantwell's name was vital to getting highly competitive NIH funding for the study. The Othmers and Cantwell designed a study in which half the kids would receive brain wave training while the other half would receive very high- and very low-frequency training as a high-tech placebo. Some critics say that the reason neurofeedback might work in some cases is merely because the kids sit still and relax. This would be a chance to see if the placebo effect was as powerful as neurofeedback itself. After the study design was worked out, Cantwell shook hands with the Othmers and told them, "I'm happy to be working with you." They submitted the study design to the Human Subjects Review Committee at UCLA, which must make sure that the experiment treats the subjects ethically. While they were waiting for approval, Cantwell died suddenly of a heart attack.

Throughout the tribulations, EEG Spectrum continued to forge ahead. Late in 1993, the Othmers had hired the first "outsider," Dennis Campbell, who had headed up a project at a small aerospace company designing advanced remote-control spy planes. Campbell had attended a seminar by Michael Hutchinson, the author of *Megabrain*, who had popularized a host of technologies, including neurofeedback, designed to enhance the brain's performance. Then someone told Campbell about the Othmers. He ended up taking their professional training course and was preparing to set up a neurofeedback practice as part of a transition out of aerospace. Just before the *Home Show* aired, Siegfried called and said he wanted Campbell to come to work for EEG Spectrum, though he didn't know how much it would pay or what, exactly, Campbell would be doing. Despite the uncertainty of the offer, Campbell recalls, "I had a feeling this was going to be big, and I said I would do it. Aerospace was coming down around my ears." His job for the first few weeks was sitting in a tiny room answering 175 to 200 letters and 40 phone calls a day as a result of the *Home Show*. "It wasn't easy," he says. "We had to tell most callers that their nearest practitioner was three states away."

Campbell's next task was to bring a new level of organization to the business. "It was such a new idea, there was nothing to base it on," he says. He structured the professional training program, created a formal marketing plan, managed sales, took care of relations with affiliates, and wrote software for some of the new protocols.

"Teaching was—and is—the heart of the Othmers approach. During the course the Othmers are center stage. Both are smart, good teachers, equally passionate about the subject but with contrasting styles. Siegfried sometimes wears a regular jacket and tie but chafes at such a uniform. He often wears a bolo tie with a bear across it and a zigzag line that looks like an EEG. He favors Birkenstock sandals and often has his collar unbuttoned. Boyishly enthusiastic, he gestures wildly with his hands during his talks, frequently laughing and cracking jokes, showing "Far Side" and other cartoons, and often making stream-of-consciousness claims and connections to other areas.

Sue, on the other hand, while she possesses a good sense of humor, is more sober, thoughtful, studious behind her glasses; and she plays the straight man when they present together. Her statements are careful and measured and carry more weight with professionals as a result. Her presentation rarely deviates from the task at hand: how to treat the brain. She is seldom at a loss for words when explaining the technology and how it can be applied: she has treated thousands of people with her own hands.

In February 2000, EEG Spectrum suffered yet another blow. The company was forced into Chapter 11 bankruptcy reorganization. The company was forced to lay off ten of its twenty-five employees, and among those who left was Dennis Campbell.

Ross Quackenbush, a Ph.D. psychologist in Salem, Oregon, is a recent convert to neurofeedback and typical of Spectrum affiliates. He has a longtime private counseling practice and has specialized in treating learning disabilities through talk therapy and behavioral interventions. He has been using neurofeedback for nearly a year. He read about the technique in a piece I wrote in *Psychology Today* in the May/June issue of 1998; by June he was attending the EEG Spectrum training course in Seattle, where we met. He was very interested but cautious about adopting the technique, for he has a good professional reputation and didn't want to jeopardize that. The first thing he did after he read the article was e-mail twenty of the best-credentialed professionals on the EEG Spectrum affiliate list. Fifteen responded with glowing comments about their involvement in neurofeedback and said it was a powerful tool that had become integral to their practice. He signed up for the course. I called Quackenbush after he had spent $12,000 on the equipment and training and a couple of months setting it up and getting it running. He told me that nine months after we first met he had no regrets about neurofeedback. "It's wonderful stuff," he said. "I've got a waiting list eight people long, and I'm ordering a second set of equipment." He is using neurofeedback with 60 percent of his patients.

To be sure, there are frustrations. A five-day training course, he said, is not nearly enough to be able to understand how to use the equipment. It only scratches the surface. "There's an absolute need for more training," he said. "I've read the manual, and there's still lots of things I don't know." Despite a $100-a-month affiliate fee, he has had trouble getting his questions about treatment for different problems answered. "I have questions all the time, and Spectrum's attitude is get on the list server and get it answered," Quackenbush said. "But I need expertise. That's what I was led to believe I would have."

One of the most often voiced criticisms of the Othmers is that the training they provide does not begin to equip people for the realities of the infinite variables that will be encountered training the human brain. Steve Rothman, who has been using the Othmers' Neurocybernetic equipment for ten years and believes they are geniuses, would like to see an overhaul in the training process. "Five days of condensed training does not begin to allow most of us to acquire more than the most rudimentary knowledge," he says. He advocates that equipment be provided only to people with master's or doctor's degrees in mental health or health care, and that they perform a more rigorous training regimen that includes several days of internship with experienced providers.

Still, says Quackenbush, neurotherapy has given him a great deal more power as a clinician. Out of fifteen people he has treated, just one has failed to respond, while the others, mostly with learning disabilities, have gained substantially. He has tested his subjects before and after and has seen gains in IQ and other scores. One young man, depressed and anxious, first tested in the ninety-ninth percentile for anxiety. After ten sessions, he was in the forty-ninth percentile: "Dead average," Quackenbush says.

His biggest miracle story, says Quackenbush, involves a ten-year-old boy with a gifted IQ whose father perished in a plane crash. The boy suffered crippling panic attacks that usually lasted for two or three hours and left him curled in the fetal position in bed. He stopped going to school because of the attacks and then stopped getting dressed—he would only wear thongs and a pair of shorts. Psychiatrists tried to treat him with antianxiety and other drugs, but he did not respond. After several sessions of neurotherapy, Quackenbush started to see changes. "I'll never forget the first time he came in wearing shoes instead of thongs. His mother said he was like his old self again, only better," Quackenbush says. After fourteen sessions the boy stopped coming because he felt he had been cured, even though Quackenbush wanted to continue treatment. It is the kind of case, Quackenbush says, that had become so physiological—that is, the stress chemicals caused by such extreme emotion had caused damage—that counseling alone couldn't resolve the conflict. "We talked some about his father's death," he says. "But it wasn't just psychological. His brain had changed. Neurofeedback changed it back again."

CHAPTER SEVEN

Paying Attention

∞

Every day on her way to work Linda Vergara witnesses the human wreckage caused by drug abuse. She drives through the urban streets of Yonkers, New York, just north of New York City, and makes her way to the Enrico Fermi School for the Performing Arts and Computer Science, where she is principal. Small knots of people gather furtively on street corners in front of boarded-up windows to exchange cash for drugs. Crack vials are scattered in the street in front of the school. Shootings are not uncommon here, and blood occasionally stains the sidewalk in front of the aging brick school when children arrive in the morning. This situation is not strange to some of the children at Fermi; many of the one thousand students come from this neighborhood and some come from families in which one or both parents have fallen victim to drugs.

Her daily drive through this small slice of despair was the first thing Vergara thought of when she was faced with a drug dilemma of her own. In 1992 officials at the private school her six-year-old son, Jon-Michael Negron, was attending, informed Vergara that he had a behavior problem. Vergara was disappointed though not terribly surprised. Jon-Michael had always been a rambunctious kid, constantly initiating physical battles with his sister or jumping up and down on the couch. He could not sit still and do simple things, such as eating or homework. He did not sleep well, and in the morning, after being awake much of the night, he had trouble waking up. "You can have a nervous breakdown when you have a child like this and don't know why," the forty-seven-year-old single mother told me. She took him to a psychologist at Eagle Hill in Connecticut, a school that specializes in learning disabilities. After three days and $1,500 worth of testing over Christmas vacation, the boy was diagnosed with attention deficit/hyperactivity disorder, or ADHD.

A child (or adult) with attention deficit disorder has a difficult time paying attention and focusing. When ADD includes the added burden of hyperactivity and impulsivity, it is called ADHD. The diagnosis of ADD has reached an epidemic proportion in recent years, and its prevalence and treatment have become extremely controversial topics. Some experts estimate that 1–3 percent of early school-aged children have full-blown ADHD, while 5–10 percent have partial ADHD with or without other problems such as depression or anxiety, and 15–20 percent show transient or subclinical symptoms. Four out of five children diagnosed with ADHD are boys. No one knows what ADD is or what causes it. Some experts believe it is a genetic problem, some say emotional, and many believe it's not a problem with the child at all but of an outdated "factory model" of education that is designed not around the learning needs of the children but around keeping order in large schools and large classrooms.

There is general agreement on what ADD looks like. A list of official symptoms includes difficulty remaining seated, acting as if driven by a motor, talking excessively, and interrupting or intruding upon others. But in a sense ADD and ADHD are in the eye of the beholder. There are no blood tests to establish who has the problem; rather, it is determined by parents and doctors, which makes the diagnosis very subjective. What one person thinks is normal rambunctiousness another might see as a behavior problem.

Vergara remembers the anguish she felt over her son's problems. "Is it genetic?" she thought. "Is it me? He's the only son I have, and he's got all these problems at age six, and he's already almost a dropout. I didn't know what I was going to do with this kid. And the only thing all of the experts say is give him a drug." The fiercest debate raging around the topic of ADD is how to treat it. Most experts recommend an integrated approach: medication, behavior therapy, and family therapy, though drugs are often relied on without the other two components. "They told me I needed to give him something to calm him down," says Vergara, an intelligent woman of Puerto Rican descent who speaks at a rapid clip. The "something" was Ritalin, a drug that successfully quiets many hyperactive children and is so prevalent in schools these days, it is sarcastically referred to as Vitamin R. When the first edition of this book was published, it was estimated that 3 percent of school-age children in the United States took Ritalin and that from 1990 to 1995 the number of kids on Ritalin tripled. At the time, 90 percent of the annual production of Ritalin—more than eight tons—was used in the United States. Since then, the use of Ritalin and similar drugs has plateaued, but is still widely prescribed.

Ritalin and Cylert and Dexedrine, two other drugs commonly prescribed for ADD, are, paradoxically, stimulants, stimulants so powerful that they are routinely bought, sold, and abused by drug addicts as an equivalent to speed. The Drug Enforcement Administration estimates that more children will be exposed to these drugs unlawfully than by prescription. In fact, injecting or selling off a child's prescription is common, especially in inner-city areas. How Ritalin works is not known specifically, but it is thought to affect two important neurochemical systems—dopamine and norepinephrine. This class of drugs has helped thousands of children who never would have been able to function in a classroom setting. In some kids, about 40 percent, the effect is dramatic and can take a child who is bouncing off the walls—talking out of turn, provoking other kids, not paying attention—and turn him or her into a quiet kid who remains seated. The drugs work at reduced effectiveness in another 25 percent of children. These drugs are not a cure, however, and work only when they are taken, usually for years at a time.

The side effects of methylphenidate (Ritalin) can be quite serious and include insomnia, a loss of appetite, a delay in growth, stomachaches and headaches, depression, anxiety, irritability, mood swings, and increased heart rate. Widespread use of the drug is so new that the effects of its prolonged use are not known, another area of controversy. Some studies indicate that children who use Ritalin are more likely to abuse drugs later in life. Nadine Lambert, a professor of education at the University of California at Berkeley, followed five hundred kids for twenty-six years and argues that the use of Ritalin makes the brain more susceptible to cocaine addiction and doubles the likelihood that a child will abuse. Another five hundred–child study by a Harvard psychiatrist named Tim Wilens and some of his colleagues argues a different side: that unmedicated ADHD leads more children to abuse stimulants later in life, ostensibly for relief.

Some critics of drug proponents say that criticizing neurofeedback as "experimental" is the pot calling the kettle black. While there are studies that show that Ritalin works well in some children, there is a profound lack of research on what ADD and ADHD are. No one knows if they are genetic, psychological, or both. There is also a disturbing lack of evidence on the long-term effects of the powerful medication. The rapid increase in use of these drugs caught the attention of the Drug Enforcement Administration in 1995. Citing a 500 percent increase in prescriptions for methylphenidate since 1990, Deputy Assistant Administrator Gene R. Haslip issued a warning:

> Let me say that medical experts agree that these drugs do help a small percentage of children who need them. But there is also strong evidence that the drugs have been greatly over-prescribed

> in some parts of the country as a panacea for behavior problems. These drugs have been over-promoted, over-marketed and over-sold, resulting in profits of some $450 million annually. This constitutes a potential health threat to many children and has also created a new source of drug abuse and illicit traffic. . . . There is a legitimate place for these drugs, but we have become the only country in the world where children are prescribed such a vast quantity of stimulants that share virtually the same properties as cocaine. We must turn down the flow that is rapidly becoming a flood.

Faced with these facts, Vergara refused Ritalin for her child. And Vergara, used to keeping order among approximately a thousand inner-city, grade school students, is firm about her decisions. "I said no," she told me one day in the chaotic hallway outside her office, as she kept watch over kids filing to their first class of the day. "I can't send that message. On the one hand we're saying, 'Take this pill, and you'll feel better.' Then when they medicate themselves with marijuana and cocaine, we scream and yell and tell them it's wrong. I wasn't going to do that. Either it's okay to medicate yourself with stimulants, or it's not." And she points out the irony of drugging kids in a school ringed by signs that read "This is a drug-free zone." At Fermi there are around a hundred kids taking Ritalin or similar drugs. Vergara had seen the effects Ritalin has on the kids in her school firsthand. While some kids are calmed, relief often comes at a cost: some were listless, had dark circles under their eyes, and seemed to have lost their passion for life.

Vergara wanted an alternative. While many parents feel the same way, they often opt for Ritalin or Cylert or Dexedrine because they are offered no alternatives and the school and teachers are pressing them to do it. But Vergara refused and, by coincidence, saw an advertisement in the newspaper that held out neurofeedback as an alternative. Vergara called Mary Jo Sabo, a psychologist who is head of a company called Biofeedback Consultants, Inc., in Suffern, New York, and who told Vergara about her experience using neurofeedback with some of her patients. Vergara was interested, though a little dubious, and decided to have Jon-Michael try it. She scheduled a series of after-school appointments with Sabo. Much to her surprise, within seven sessions, Vergara says, she started noticing changes. Morning was less of an ordeal. "I used to have to ask him to get dressed twenty times," says Vergara. "That was changing. He was getting up in the morning without crying and fighting." Jon-Michael would sit longer at the dinner table. And he spent more time on his homework. "One day he came home and sat down and started doing his homework without being asked," she says. "I almost fainted." As was the case with the Othmers, she wondered if she wasn't just imagining things were getting better because she wanted them to. But the improvements continued. Disbelief turned to amazement, and her reaction was the same as the Othmers': Why wasn't this out there? Why didn't more people know about this? She committed herself to what would become a controversial project: teaching grade school students to train their brain instead of using drugs.

One of the experts Vergara later turned to for the project at Yonkers was Dr. Joel Lubar, the country's foremost authority on the treatment of ADD and ADHD with neurofeedback, having studied and treated the disorder in both research and clinical settings for more than thirty years.

Lubar started college at the University of Chicago in 1957, intending to become an astrophysicist. In his junior year he took a couple of courses in a new multidisciplinary program the school had just started called biopsychology, the study of how emotions affect human physiology. "I took these courses, and I thought, 'The brain is even more interesting than astrophysics,'" he says, and he changed majors. When he graduated, there were no good interdisciplinary programs anywhere else, so he stayed in Chicago. He crammed eight years of work into six, and in 1963—at the age of 24—received his Ph.D. in biopsychology, one of the first advanced degrees of its kind in the country. He taught for a few years at the University of Rochester, where he was younger than many of his students. Then the University of Tennessee in Knoxville recruited him in 1967, promising him a laboratory that covered an entire floor to conduct his research. By the age of twenty-nine he was a full professor. He has been at the University of Tennessee ever since, teaching several courses, including behavioral medicine, neuroanatomy, physiological psychology, and quantitative electroencephalography, or QEEG, a more complex way of measuring the brain's electrical activity than a simple EEG. He has nine graduate students working with him. In 1971 he was a visiting professor at the University of Norway in Bergen, the main medical school in that country. Since 1980 Lubar has worked closely with his wife, Judith, who has a master's degree in clinical social work as well as a bachelor of arts in medieval history and a master's in foreign languages.

Lubar's focus for the last four decades, in a nutshell, has been the study of how the physiology of the brain—the cells and blood supply and physical characteristics of the brain—affects how people think and feel, essentially how brain creates mind. Intelligence, for example, is governed in part by the density of the neuropile, the collection of cells in the brain, he says. Another key factor in intelligence is how quickly the brain can allocate and reallocate blood supply, which activates parts of the brain needed for a specific task. People who are more intelligent can get blood quickly to the part of the brain where it is needed and, just as important, stop the flow quickly and shuttle blood elsewhere.

In 1967 Lubar was doing basic "cat work" in the graduate lab at the University of Chicago, researching the behavioral effects of physical lesions on the brain. In one experiment, he wired a metal food dish so that when a food-deprived cat ate from it, "the cat would get a mild electric shock right through the mouth," he says. "They would leap back and their hair would stand on end, and typically when that happened, the animal would not go near that dish ever again in that setting." But there were exceptions, created by the scientist. Lubar had performed neurosurgery on some of the cats, opening their skulls and suctioning a tiny lesion into the septal area, which is in the forebrain. Despite the shock, those cats with the lesion, and only those cats, returned to the food bowl again and again. "They would pant, they would vocalize, like the cats without lesions, but they kept responding, kept coming back to the food bowl. It's like a hyperactive child who does the same thing over and over again." Why? "Because this area of the brain controls the ability of the motor cortex to shut down when there has been punishment," Lubar says. (The septal area also affects emotions, Lubar says, and the cats became very affectionate, "like puppy dogs," he says. "They follow you all over the place.") The septal area is called the subcallosol cortex in humans, a small, grape-sized piece of tissue under the front of the corpus callosum, the strip of brain that joins the two hemispheres. Though small, it is a key "crossroads" region of the brain, which funnels communication in the form of three key neurotransmitters—dopamine, norepinephrine, and serotonin—from the limbic system, motor cortex, and amygdala to the prefrontal lobes. The prefrontal lobe of the brain is critical to how we act as social beings—the so-called organ of civilization. It governs such things as planning for the future and learning from mistakes. Essentially, the prefrontal cortex accepts input from the emotional control areas, the sensory and motor areas, and other regions and integrates them all into coherent behavior. A hyperactive child, Lubar hypothesizes, may have a smaller or weakly functioning subcallosol cortex, or one that has a poor blood supply, and so the prefrontal cortex is not fully engaged and the mitigating or controlling effect on behavior does not happen.

The term *attention deficit /hyperactivity disorder* was decades off, but the notion of "a deficit in behavior control," or what was called hyperkinetic syndrome, goes back to 1905, when George Frederic Still described it in *Lancet*. There are some children, Still wrote, "who are characterized by wanton destructiveness and a deficit in moral behavior." Lubar's early work demonstrated that uncontrolled behavior may be based in physiology and not a lack of will or a bad personality. "ADHD is a neurobiological disorder," says Lubar. "The subcallosol cortex is key as to whether behavior is appropriate or not."

Lubar expanded his inquiry, and the next study had three populations of cats. One had a human-made lesion on the septal area, as before. Another group had a human-made lesion on the cingulate gyrus, a part of the brain that controls an animal's ability to inhibit responses. The inhibition in these animals was increased, and they became fearful and very edgy—the very opposite of those animals that were hyperactive. When lesions were created on both areas in the third population, the effects canceled each other out.

In 1972, as Lubar was busily suctioning brain lesions in cats, Sterman's first paper on training epileptics with neurofeedback was published in the journal *EEG and Clinical Neurophysiology.* Lubar read it and was floored. Right away, he says, he was excited because he knew there was an application to his work. "Sterman's paper said there is this rhythm, the sensorimotor rhythm or SMR, that may be dysfunctional in epileptics, and when the rhythm is partially restored, a resistance to seizures occurs. As soon as I read that, I said, 'My God, I think this will work for controlling hyperactivity in children because the circuitry is very similar.'" If you could quiet motor responses for epilepsy, he reasoned, quieting motor responses in hyperactive children should be a piece of cake. Lubar and a colleague, Bill Bahler, and a graduate student, Ron Seifert, were the first of several to replicate independently Sterman's epilepsy work. Then Lubar applied for, and received, a fellowship from the NSF to spend an academic year learning the neurofeedback technique from Sterman himself. In 1976 he packed up Judith and their three children and moved to Los Angeles so he could work with Sterman.

Ever since, Lubar's work has been primarily about ADD and ADHD, and in the field of neurofeedback he is considered the reigning expert. He has written more than one hundred papers and conducted more than twenty-five studies of the technique. In 1975 his work with Ron Seifert built on Sterman's SMR protocol by teaching the client to inhibit theta, which is now part of nearly every protocol. One of the most powerful studies was an A-B-A design in 1975. Four children were trained until their ADHD disappeared and then trained in the opposite way until tests showed it was back. Then they were trained again until tests and EEGs showed that they were asymptomatic. "It worked beautifully," Lubar says. All of Lubar's work has had similar precise and carefully controlled designs using statistical measures.

Lubar's hypothesis for what is happening in an ADD brain is that with certain subtypes of attention deficits, there is a decreased metabolism and decreased blood flow to the subcallosol cortex. "They're turned off," he says. "That means those areas are not getting enough norepinephrine and other neurotransmitters." As is the case with epilepsy, Lubar believes that when the brain is trained with neurofeedback, blood bathes the cells in the frontal cortex and acts as a kind of fertilizer helping cells overcome malformation, due either to genetics or perhaps from cortisol damage caused by emotional stress. Existing connections are strengthened or reorganized, or perhaps they grow new branches. Whatever the case, they make better, more robust connections with adjoining cells, and so the transfer of current and neurochemicals works much faster and more efficiently. "What we hope we are doing is turning that area on," he says. However, Lubar doesn't think that the neurofeedback is treating the subcallosol cortex directly, because it's deep in the brain. "We're training areas in the central and frontal cortex that are involved in the focusing of attention, intent to attend, and the evaluation of appropriate behavior," he says. "By training the area that the subcallosol cortex projects or connects to, it reflects back on how it is operating. That's what we hypothesize."

Typically, a Lubar client does forty sessions, each thirty to forty-five minutes long, for ADD without hyperactivity. For ADHD, it typically takes ten to fifteen additional sessions. The total cost of a full evaluation and treatment ranges between $3,000 and $4,000—"which is cheaper," Lubar says, "than taking Ritalin for five years." (Prices vary, but a five-year prescription for Ritalin costs about $3,600 for the drug alone, not including doctor fees.) Lubar uses a computerized biofeedback system called the A-620, which he designed for Autogenics, a division of Stoelting Company, the Chicago–area electronics firm that is one of the first and largest biofeedback companies. The A-620 and the Othmers/Ayers instrument came out around the same time and were the first generation of computerized instruments. The Autogenics equipment is similar to the Othmers' equipment in that it uses game performance to reward the client, featuring an elaborate space game that includes fetching fuel pods and racing other ships, as well as hundreds of puzzles. It is a highly regarded instrument in the field. Lubar's protocol, inhibiting both 4–8-hertz slow waves and 6–10-hertz waves, and encouraging activity in the 16–20-hertz range, has been adopted by most of the rest of the field.

Lubar believes that neurofeedback should always be combined with family and academic counseling, which his wife, Judith, supplies. "Whatever happens in neurofeedback happens in the context of the family," says Judith Lubar. Biofeedback isn't done *to* a person, it's something they must want to learn. A key part of the process is to engage the will. Many families do not agree that the child has a problem, how severe it is, and what to do. If everyone comes in angry and upset, the child is not going to learn." Training also restructures family dynamics and counseling helps with that. "One individual is going to be changing," she says. "A lot of other things are going to change."

Lubar's clinic treats about 70 percent of the people who are referred there, exempting people with depression and personality disorders. Of the population he treats, Lubar says, better than 90 percent respond very well to neurotherapy. After the training, people think more clearly, pay attention much better, and get better grades in school. Lubar also makes the controversial claim that the people he treats have seen their IQ rise by an average of eleven points. "I think it's largely because they're more motivated," he says. Despite the improvements routinely seen, and despite the fact that most people become what Lubar calls "asymptomatic," he insists that neurofeedback does not normalize the brain. "We never use the word *cure,*" he says. "Never. Cure means not only no symptoms, but also that the underlying problem is also gone. That's not the case." Under stress a problem may return, albeit in a lesser way, and a booster session or two may be required, Lubar says. Sterman has seen the same thing. A woman who had been seizure-free for years called and told Sterman she was going through a divorce and had a feeling her seizures could recur. Could she do another session? She did, and increased her seizure threshold back to where it had been. Boosters are usually just a session or two.

Lubar conducts his own professional training course, which he has been offering since the 1980s to medical and mental health professionals. Some three hundred people have been through his course.

The one thing that has eluded Lubar, however, is a double-blind, controlled study, the gold standard of research that could earn him and his technique the place in science he feels is his due. He needs such a study, not to convince himself, he says, but to convince a handful of vociferous critics who have relentlessly pointed out his lack of double-blind results.

There are several reasons that, except for Sterman's epilepsy studies, neurofeedback has not had that kind of research, he says. The big one is funding. A true double-blind study could take years, as it did with Sterman, and could cost hundreds of thousands or millions of dollars. Nearly all practitioners of neurofeedback are small operations, not large companies. There are no huge, multi-billion dollar companies out there that can afford to bankroll one of these studies, biofeedback practitioners say. Second, it is difficult to do double-blind, controlled studies with neurofeedback. No one knows what is in a pill, but both practitioner and patient can often quickly figure out whether a patient is getting the real biofeedback training or the sham.

Some who practice neurofeedback don't believe brain wave training needs to meet the same standards required for drug testing. Lynda Thompson, a neurofeedback practitioner in Toronto, is among those. "Everyone accepts research that shows there are cardiac benefits from exercise," said Thompson. "Those are not double-blind studies, they are outcome. I think that's what we should accept as valid with neurofeedback rather than the drug model. It's unfair to hold us to the same standard as drug research." In fact, the Office of Technology Assessment, an agency of the U.S. Congress, estimates that less than 30 percent of current conventional medical procedures have been thoroughly tested with double-blind, controlled randomized studies. Most surgeries, for example, from heart bypass to tonsillectomies, are done without such studies; they are based on outcome. Sometimes drugs are used even when they have not been proven effective. "Thirteen published double-blind placebo-controlled clinical trials of antidepressants for the treatment of adolescents failed to demonstrate an advantage for active drug over placebo," wrote Dr. Leon Eisenberg of the Harvard Medical School in the *American Psychological Association Journal* in 1999. "Yet physicians wrote some 4 to 6 million prescriptions in 1992 for patients 18 and under." There are double-blind studies showing the efficacy for adults but, as is the case with Ritalin, drugs can have a very different effect on children and adults. In fact, prescription drugs are widely prescribed by doctors for applications for which they were not originally intended or studied, something called off-label use. Anticonvulsants, for example, which were designed to control seizures, are being used to treat panic attacks, migraines, schizophrenia, and even rage. And there is electroconvulsive therapy, the shock therapy of *One Flew Over the Cuckoo's Nest* fame. (Sterman refers to electroshock's gross effect on the brain as "kicking the TV" or "pressing the reset button" and can't believe people will resort to that before they consider neurofeedback.) Abandoned because it was perceived as barbaric, it has become kinder and gentler and is considered very effective for some kinds of difficult-to-treat depression. But it is still an indiscriminate shot of electricity through the unimaginably delicate, microscopic circuitry of the brain. And there are no double-blind, controlled studies of its effectiveness. When doctors prescribe a treatment that no or few studies support, because they see that it works, that is based on something called clinical judgment. Lubar and others feel they are being held to a very high standard of proof, not for scientific reasons, but because of an irrational prejudice against biofeedback.

Another part of the problem is the fact that biofeedback doesn't fit neatly into any category. Sterman's and Lubar's work with epilepsy and ADD has cast neurofeedback as a medical intervention, an alternative to drugs, and that is how the technique is often touted. "If you make that claim, then you are bound to make the same justification as someone who is using drugs," says Chris Carroll. "The FDA would not approve a medication without a double-blind, controlled study. And they wouldn't approve this. That's the reality we have to face. But there's another perspective. If it's viewed as more psychological than medical, as a cognitive or behavioral therapy, rather than sold as a substitute for Ritalin, then you use the psychotherapeutic standards. There you find acceptability."

The role of the FDA in all this is a little uncertain. New biofeedback equipment is grandfathered under a 1976 law that regulates the older biofeedback equipment, something called a 5-10K exemption, though that law did not foresee anything as unusual as beta and SMR training. That law allowed biofeedback to be used for general relaxation and muscle reeducation, but manufacturers cannot make claims of successful treatment for such things as attention deficit disorder and depression and closed head injury. (Clinicians can lawfully make those claims, however.)

The FDA oversees only manufacture, not how biofeedback equipment is used or the people who use it. "We don't regulate the practice of medicine, and biofeedback devices are treated like any other medical device," said a spokeswoman for the federal agency who asked not to be identified. "Our goal is to make sure the product is safe and effective for its intended use." If manufacturers were to claim effective treatment for ADD or epilepsy "or any specific medical claim, we would want to see clinical studies on safety and effectiveness. That's true of any medical device." Manufacturers, such as the Othmers' company, Neurocybernetics, or Autogenics or Lexicor (another neurofeedback instrument manufacturer), specifically do not make claims for treatment of any medical conditions, though the equipment is widely used for such treatment.

Although the process is done very openly, in a sense, neurofeedback manufacturers are flying under the radar of the FDA, and many worry about a crackdown. One of the reasons there hasn't been much of a brouhaha about neurofeedback by regulators is that most people don't believe it works, but practitioners fear that as the power of the technique becomes apparent, the industry will be more at risk from regulators. An all-out war with the FDA could ruin providers operating on a thin profit margin. Or the technique could be "medicalized"—that is, the FDA could rule that only doctors or other trained professionals are allowed to administer brain wave training. Because negative side effects are rare, the Othmers and others believe that everyone should have access to neurofeedback, either in the office of a practitioner or in their home, to train themselves to treat problems or to enhance already normal functioning.

In some ways, it doesn't matter all that much if there are no double-blind, controlled randomized studies. Parents of children with ADD and ADHD, especially well-educated, middle-class parents who detest the notion of drugging children and are concerned about the unknown long-term use of stimulants, are propelling the market for neurofeedback, more than any other single group. There is a strong antidrug backlash, especially given the sheer number of kids who are suddenly taking stimulants and several well-publicized drug disasters, such as that of the diet drug phen-fen, in which people died after the drug had been approved by the FDA. Canadian officials banned the ADD drug

Cylert because it is believed to cause liver damage and was implicated in the death of a young man. A study published in 1998 in the *Journal of the American Medical Association* showed that 106,000 hospital patients die and 2.2 million are injured each year by adverse reactions to prescription drugs. And experts say that as the number of medications grows, so does the problem. The lack of drug-style research does reduce the willingness of third-party payers to pay for the service, however. Lubar estimates that half of all insurance companies—others estimate less—will pay for biofeedback, and that limits the technique largely to parents who have the resources to pay for it.

After weighing the options, it was, says Linda Vergara, a no-brainer for her to move neurofeedback into the Fermi School for what she refers to as "her babies." Her own son was no longer hyperactive, though he is still a little ADD. "When I tell teachers he was hyperactive, they can't believe it. He is very mellow. He's not aggressive. Not violent. I thought I was going to have an aggressive kid on my hands." There was so much to gain, she says, and relatively little to lose. Not only for her own students, but for children around the country. Ritalin and other medications were the lesser of two evils for many kids, she felt. But now there was an option that did not present harmful effects at all. She went to an administrator, a friend of hers, who said she would support Vergara's plan but could not commit any funding. So Vergara wrote a grant proposal to a state program called Limited English Proficiency, which exists to help teach English to Spanish-speaking children, and was awarded $20,000. She bought three new computers for the English-language classroom and spent $2,500 on a fourth computer, a neurofeedback system called Cap Scan, built by American Biotec, a company in Ossining, New York. Administrators of the grant money called and said they could not approve spending on something that appeared to have nothing to do with language. Vergara did what she felt she had to do. "I lied," she says. "I told them it was language software and they had to approve it. And they did." She sent a guidance counselor, a psychologist, and a teacher's aide to Mary Jo Sabo's to get training. They set the machine up in a small room near the principal's office.

One of the first things Vergara and Sabo did was to set up a small study designed by Dennis Carmody, a clinical psychologist at St. Peter's College in New Jersey. Eight kids were in the experimental group, which received twenty to twenty-five minutes a day of neurofeedback. Eight others in a control group received no training. The groups were selected at random, though they had an equal number of boys and girls. Before the beginning and at the end of forty sessions, says Vergara, she and other observers went into classrooms and recorded what they saw going on, on a special grid that charted behavior. "We found drastic changes: Increased attention spans. On task for longer periods of time. Less disruptive behavior. Fewer outbursts." Suspension rates, attendance, and tardiness all improved for the experimental group. Training at Fermi was the same as elsewhere: inhibit or discourage the slow waves and encourage the waves that are indicative of paying attention.

Though the study was an informal one, parents at the school were enthusiastic. None was more so than Faten Hussein, whose son Mohammed was something of a little criminal. At first she was not in favor of the training. But at her wit's end, she relented and soon became one of its biggest supporters. "My life was miserable," she says, in rapid-fire English with a thick Egyptian accent. "The miserable thing in our family? Mohammed. I ask God, why did you give him to me? He was very bad. He could not focus. He was very bad. He could never sit and do something. His head was very confusing." Mohammed had a traumatic childhood and had often banged his head against his crib. He was known as a bully at school, punching kids and attacking and choking a cousin. He was given fifty-eight sessions of neurofeedback in 1997. "Now he's like an absolutely normal kid," says his mother. "He has problems, but when he says 'I'm sorry,' he means it. I'm saying, What is this neurofeedback? It's like magic, what happened to

Mohammed." Mrs. Hussein became a crusader for the technique, vigorously lobbying local officials for more funding.

Mohammed feels the same way. "Biofeedback is something extraordinary good," the pudgy ten year old told me, waving his hands in the air for emphasis. "They're not making me say this. It really feels so good. I used to be the biggest, baddest bully in the school. I'd bully kids bigger than me. Now I'm different. It calms you down. It makes you feel good." Mohammed says he looked forward to his sessions because after he was finished, he felt more relaxed.

ADD and ADHD aren't the only things that changed at Fermi. Rachel Campanella is a fourth-grade teacher. She had a young man in her class named Nelson Mercado, who did not fit into the social milieu at all. He was never a discipline problem; he simply kept to himself. He had no self-esteem, no confidence. His language skills were that of a kindergarten boy. He kept his head down on his desk all day. Neurofeedback sessions started for him in March of 1997. "By June I saw major changes," Campanella says. "He spoke, he smiled, he raised his hand and spoke in sentences. I am an advocate because I saw what it did for a boy who was so unhappy. I never knew if he had teeth, because he never smiled. Now he comes by with a big grin and says, 'Good morning, Mrs. Campanella.'"

There are children for whom neurofeedback is not enough—the ones who leave school every day and return to a chaotic, sometimes violent, home, or a home where they are neglected. The school tries to ease the tempestuous external pressures on many of the kids in a district where 91 percent of the families live at or below the poverty level, by offering wholesome breakfasts, lunches, and dinners; parenting classes; job training; and counseling. But there is only so much that can be done at school. One boy the school tried to help was a real discipline problem. His mother was a crack addict and his father a junkie. He lived at home with his grandmother and her boyfriend, who were both alcoholics and who sexually molested him. His behavior was bizarre, including wetting the bed every night and urinating in cans that he kept his bedroom. "We stopped the problems with urination," Vergara says. "but I don't know about his other problems. We'll have to see."

Yonkers school officials liked what they saw at Fermi. The school has been allocated resources to buy a total of ten neurofeedback instruments, and two other schools, P.S. 9 and P.S. 13, have purchased the equipment. The district-wide budget has grown from $2,500 the first year to $143,000. Rafael Diaz, a Yonkers school trustee, has been a staunch advocate of neurofeedback training and sees it as an issue of empowerment. "Minority children are more likely to be labeled a behavior problem and be drugged," he says. "Their parents are more likely to accept what a doctor tells them. I could see what was going on in our school. A lot of our students are being given no alternative to drugs. We should not be drugging our children unless there is no viable alternative. And biofeedback is a viable alternative." The Yonkers project ended after the attacks of 9/11, when funding for many of the nonessential programs at New York schools were diverted to deal with the aftermath of the attacks.

The difficulty for a journalist covering the field of neurofeedback is that it is so recent, and so unlike any other existing therapies, that there are few informed critics. Many experts hear the term "biofeedback," roll their eyes, and immediately lump it into the same category as reading auras and balancing chakras. It is easy to understand why experts would be critical of neurofeedback without closely examining it. They've spent twenty or thirty years in their fields and have become well known and respected as a source of information. Along comes a therapy that is so far out of the box they may never have heard of, let alone examined, it, and it purports to be far better than the therapies they are expert in.

Plenty of people, many with topflight credentials, who know nothing about neurofeedback are more than willing to denounce it as impossible. I have chosen not to quote professionals in that category for obvious reasons. I have not found anyone who has used the technique and feels it is hogwash (though some who use it feel some of the claims are overblown). I have found people who have read some of the studies and who are critical of neurofeedback. One of those is Russell Barkley.

To many in the field of neurofeedback, Dr. Russell A. Barkley is the prince of darkness. At every neurofeedback conference someone brings up Dr. Barkley's name, and it is a surefire way to rally the forces against a common enemy. A Ph.D. psychologist at the University of Massachusetts Medical School, Barkley is one of the country's leading experts on ADD and ADHD. He has written a long list of books on the subject, including *ADHD in Adolescents: Diagnosis and Treatment* and *Defiant Children: A Clinician's Manual for Assessment and Family Intervention.* He has also conducted a number of studies for drug companies and is a leading proponent of the use of Ritalin and other drugs, in conjunction with behavioral and family therapy, to treat ADHD. He has vociferously denounced neurofeedback as experimental and unproven. In an interview, he told me that he had "read some papers a few years back" on neurofeedback but had not closely examined the technique and did not use the therapy himself. He was not familiar with Sterman's work. "We don't have any studies that say it's bad for you, and I don't think it will do harm," Dr. Barkley said. "But I don't think it should replace treatments that are cheaper. Claims that it is as good or better than medication are totally unfounded. Claims that it alters brain physiology are totally unfounded and unethical."

And that, of course, is what precisely many neurofeedback researchers, including Sterman and Lubar, do claim. They believe that, at a minimum, neurofeedback training brings about a functional reorganization in the brain and makes the connections between cells more robust.

Barkley, echoing what many critics believe, suspects what is really at work is a placebo effect. "Case studies prove nothing because they're totally uncontrolled. There's an aura of medical intervention here. High technology in a medical environment has a high placebo effect," he says.

"It's not the equipment. It's the exercises, the mental exercises they are telling these kids to do." Some kids may be getting better "with maturation alone," he says. "And some kids don't have ADD." Barkley is critical of Lubar's studies because of their small sample size and lack of a double blind. "Parents need to be fully informed about the lack of research," he says. "Basically, it's buyer beware."

Another outspoken critic of neurofeedback is Sam Goldstein, an assistant professor of neuropsychology at the University of Utah in Salt Lake City who also sees patients at a private clinic. He also has written several books on the topic, including *Attention Deficit Disorder and Learning Disabilities: Reality, Myths and Controversial Treatments.*

Goldstein does not say brain wave training doesn't work. He believes there isn't enough data to decide whether it works. "The older Lubar studies are not well done," he says. "Subject selection was clearly a problem. He just took people who showed up"—a good study needs a random selection of subjects. And, he says, there was hardly any follow-up on how subjects fared over the long term.

Miracle stories, he says, echoing Barkley, don't cut it. "They are seductive. It's a correlation, but it doesn't prove cause and effect. There are a number of possible explanations. One is, it works. Two, it's a placebo. Three is, there's some mechanism operating that we don't understand." Without definitive research there is a temptation to stretch claims, he says. "Look at all the key guys in the field," Goldstein says. "Their opinion is integrally tied to their ability to make a living. If Sterman comes out and says it doesn't work, he's out of a job. If Lubar comes out and says it doesn't work, he'll have to go back to academia. If Othmer says it doesn't work, his business goes down the tubes. They have a great deal of interest in saying, 'Look how wonderful this is.' I would like these guys to do more research, and I support their application for research grants."

While some critics have lashed out particularly at Lubar's claim of higher IQ scores in those who have completed the training, Goldstein says that if neurofeedback does work, the rise in IQ score is no big deal. "It's reasonable to assume that the same increase you would see in medicine with kids who take IQ tests—about ten points—is the same you would see with biofeedback. It's because they're performing better, not because they're smarter."

Barry Sterman allows that the effects of neurofeedback may well be from the placebo effect, at least in part. In fact, he believes neurofeedback may be tapping the same mechanism that the placebo taps, the same capacity to generate a change in the brain to heal. No one knows, because placebo effect has not been well studied. But if it is placebo, he says, that is only part of the equation. "There is very real, long-lasting, physical change in the brain," he says. "The first sign that something is happening is that the QEEG shows that the brain waves are changing in the direction I am shaping them." Most importantly, he said, is that the changes last.

Despite a lack of funding, serious science is starting to happen, especially in the field of ADHD, to build on the work that Sterman and Lubar began. One of the best studies to date was published during the summer of 1999 in *Neuropsychology,* a respected scientific publication published by the American Psychological Association. Vincent Monastra, a Ph.D. research psychologist and family therapist in Endicott, New York, who is also a neurofeedback practitioner and collaborator of Joel Lubar's, organized a study to look at the role theta plays in ADHD. Eight investigators collected EEGs at a single site—the top center, or CZ—in 482 subjects. (It was replicated later in another 440 subjects.) They averaged the theta and beta readings, to come up with a ratio—how much time was spent in the spacey theta state versus the alert beta waves. Findings in six- to eleven-year-olds were most dramatic. Children who had been diagnosed with inattentive ADD and ADHD by physicians and parents using the traditional diagnostic criteria, each had about nine parts theta to one part beta. Controls, on the other hand, had three parts theta to one part beta, a third of the slow wave activity. "It would be like listening to a radio transmission or talking on a phone," said Monastra. "And having the signal cut out three times as much as the average person." The dramatic difference in an ADHD brain and a normal brain becomes very clear, he said, when you hook up a child with ADD to an EEG and "give them something to read and within ten seconds their EEG drops into theta and it looks like they are in the middle of a daydream."

The study doesn't prove the efficacy of neurofeedback, but the theta-beta ratio establishes a strong, accurate, and objective way—as opposed to subjective parent reports or a less accurate TOVA—to use EEG to diagnose ADHD and to measure the before and after effects of brain wave training. Monastra is now working on a study to find a similar EEG fingerprint for depression, conduct disorder, oppositional defiance, and anxiety.

In December 2002 Monastra published what is widely considered one of the most definitive studies of neurotherapy, in the Journal of Applied Psychophysiology and Biofeedback. A hundred subjects participated, ages 6 to 19, all of them taking stimulant medication. Everyone was tested at the start. During the study half of the population received neurofeedback [based on Lubar's protocol's] half did not. After a year all subjects were tested again. Only those who received neurofeedback were able sustain their gain in performance, measured by TOVA and EEG, when tested off their medication.

The idea of a normal database is one that will play a role in the future of EEG biofeedback, and is the way the field is heading. A patient will come in for a QEEG, and this fingerprint of their brain wave activity will be compared against a host of normals. A clinician will know immediately how the brain wave activity needs to be shaped—which sites need to be trained up or down. Libraries of EEG fingerprints already exist—Dr. E. Roy John, a brain researcher at NYU, has collected thousands of QEEGs from around the world and commercialized them for clinical use.

Another criticism is that the research in the field of neurofeedback is old and fails to take advantage of new technologies, such as the functional MRI, PET scans, and 128-channel EEGs. Those technologies could answer many questions about neurofeedback and provide a much more detailed picture than a 19-channel EEG of how much the brain is changing. Richard Davidson is a respected Ph.D. research psychologist at the University of Wisconsin at Madison who does brain imaging in a state-of-the-art laboratory. He is very familiar with neurofeedback. "The standards for quality research in the field [of biofeedback] are not good," says Davidson. "It's not up to what it should be. Serious neuroscientists would think most of the research is crap. It doesn't take into account sophisticated advances in EEG and brain imaging. That's not to say there isn't an important grain of truth there. There is. There's enough there to warrant serious research."

In 2006, after this book was first published a landmark study was conducted using state-of-the-art brain imaging technology at the University of Montreal. It should go a long way toward silencing critics of the science of neurofeedback. Conducted by principle investigator Dr. Mario Beauregard, the study found a significant difference in children with ADHD who had brainwave training and those who did not. Children were scanned with fMRI one week before and one week after neurofeedback training. Results, published both in Neuroscience Letters and Applied Psychophysiology and Biofeedback found significant changes in the two groups. The study concluded "there is mounting evidence that neurofeedback training (NFT) can significantly improve cognitive functioning in AD/HD children." The training, Dr. Beauregard said, normalizes the functioning of a part of the brain called the anterior cingulate cortex, the ACC, a key player in how we pay attention.

In some ways the lack of funding is a catch-22 situation. Without a large body of solid research, it's hard to get funding; and it's hard to build a body of research without that funding. The inability to get funding reflects on biofeedback's reputation, as well as the politics of federal research grants. "The federal government follows the lead of the powers that be when handing out research money," says Chris

Carroll. "You need to be an established international authority with proper credentials. If someone submits a grant for ADD research and neurofeedback, for example, they need to be an ADD expert and not an expert in neurofeedback. And ADD experts believe in medication as the treatment of choice."

Dr. Alan Strohmayer, the former chief of Biofeedback Services at North Shore University Hospital in Manhasset, New York, was asked by his boss to take a look at the new brain wave biofeedback when it was new. He was very skeptical about brain wave training, he says, but he read the scant scientific literature and told his boss there wasn't much to support the claims. Lubar's studies had appeared only in small biofeedback journals. His boss asked him to look further, and he went to an AAPB conference in Denver. "I heard all this hype—'We can cure this, we can cure that—and I looked at it and I said exactly what many of the critics would say, 'This is insane. It sounds nuts,'" Strohmayer recalls. "People were enthusiastic, evangelical, but there was no data there." When a little more data came in, Strohmayer decided to give it a try and started using it at the hospital. And the results, he says, have been powerful. "I got the same phenomena," he says. "For example, I had a Tourette's kid jumping up and down in the chair, and halfway through the game he would be sitting still. And I did little tricks to see if it was placebo, like unplugging the leads to the equipment without telling them, and they would start jumping up and down again, so I knew it was really working. Getting the phenomena in your own laboratory is what counts." Many of the Lubar studies, he agrees, have flaws. They weren't carefully randomized. There wasn't much follow-up. They used small sample sizes. But the before-and-after testing with the QEEG and other methods is careful and numerous. The changes are predictable across the board, and unlike most placebo effects, the changes last. Any scientist who doesn't want to agree with another can find a flaw in a study if he looks hard enough, Strohmayer says. "The bottom line is that Lubar is as honest as the day is long," he says. "He wouldn't make anything up. He's careful. Lubar demonstrated efficacy, and that it works in a therapeutic way." Research done by others—though short of double-blind, controlled studies—also shows that neurofeedback works, he says.

"There's something else at work" in the criticism of neurofeedback, Strohmayer believes. "As medical spending is cut back, there's a turf war for health care dollars. Anybody who is not traditionally trained and licensed is being persecuted. Neurofeedback is a major threat to the medical industrial complex. It doesn't fit in. It's not something they can package and sell, like drugs, and make large amounts of money." Where does it leave traditional medicine, he wonders, if a majority of the problems they treat are stress-related and an effective alternative approach to treat stress is found? "The stakes are high here," he says.

Dr. Jonathan Walker, the neurologist in Texas who uses neurofeedback to train closed head injuries, headaches, and other problems, says he knows why neurofeedback isn't more widely available. The cost of an anterior lobectomy, a surgical procedure that removes the small bit of tissue in the brain that is causing epileptic seizures, he says, generates about $200,000 for the neurologist, surgeon, and hospital. "Where's the incentive to use neurofeedback?" he said.

Proponents say that neurofeedback's role in the treatment of ADHD may extend beyond merely calming and focusing kids in school. A growing body of research indicates that ADHD kids are far more likely to become criminals and thrill seekers. Because of slower than normal waves in the prefrontal cortex, which governs our planning and social relationships and organizes behavior, the front of the brain cannot communicate effectively with the midbrain and properly regulate emotions in everyday life. Remember the metal spike that pierced the frontal lobe of Phineas Gage and turned him from a hard-working, polite individual into a foul-mouthed ne'er-do-well? There's a similar mechanism at work. "It leads to an inability to evaluate appropriate behavior in a social situation," says Lubar. "They live for the moment. They focus on immediate gratification."

Damage to the cortex—from accidents, or drinking by the mother, abuse, genetics, or oxygen deprivation at birth—apparently causes slower than normal waves in the front of the brain and the inability to engage the other systems. With the emotional center of the brain not appropriately engaged in the world, these children and adults are without the usual levels of fear, remorse, and sadness. This can explain why many children with ADHD do not behave well socially.

Because the brain's internal communication system in people with slow wave activity in the frontal lobes is sluggish, incoming messages need to be more sensational than normal to achieve the same level of stimulation. Not everyone with a preponderance of slow waves in the cortex grows up to be a criminal. There are usually other risk factors, such as parenting and the social environment. And the craving for risk and adventure can be the catalyst for a successful career or can lead someone into thrilling, high-risk sports such as mountain climbing and skydiving. But a far larger proportion of people with ADHD will be arrested for felonies. Adrian Raine, the psychologist at the University of Southern California, has studied arousal levels in juvenile delinquents. In one study he and his associates followed 101 school kids from age fifteen to age twenty-nine and studied their resting heart rate, their skin conductance, and their EEG, which measures central nervous system arousal. What they found is that kids with low arousal in the prefrontal cortex had a much greater chance of becoming delinquents. The theory is that these kids become addicted to stimulation. "Kids with low arousal seek out stimulation to increase their arousal levels back to normal," Dr. Raine says. "For some adolescents joining a gang, burgling a house, or beating someone up is a way of getting an arousal jag."

The Othmers, Lubar, and other neurofeedback proponents say that brain wave training, because it treats ADHD, reduces a client's propensity toward criminal behavior. And there are a number of case reports—though no studies—of young people, mostly men, who were headed down a path of delinquency and after neurofeedback training were able to make better, less impulsive choices and turn their lives around. One of those was a young man named Vern, who as a boy had ADHD. As he became a teenager, he started using drugs and stealing constantly, even though the family had plenty of money. His mother heard about neurotherapy and, at her wit's end, started him on three sessions per week. "He went from Fs to As in three weeks," his mother told me. "He could remember his lunch money and remember his books." He lost interest, she said, in hanging out on the streets. "He was focused and could carry on a conversation without being distracted. He was interesting. For the first time I was getting to know this kid." Vern now has a good job, and his criminal behavior is a thing of the past.

Dr. Raine is intrigued by the use of neurofeedback for such conduct disorders. "Case studies are a start," he says, "but there have been no systematic scientific studies in the reduction of conduct disorder to show that biofeedback training works."

As an increasing number of people turn to neurofeedback to solve problems that drugs or other therapies cannot treat or treat poorly, scientific studies may be rendered irrelevant. As the exponential growth in herbal medicines and alternative treatments shows, the health care system is first and foremost market-driven. If the reports about neurofeedback from parents and schools continue to be positive, that may be all that matters.

CHAPTER EIGHT

A Return to Deep States

∞

In 1990, biofeedback practitioners at the annual meeting of the Association of Applied Psychophysiology and Biofeedback in Washington, D.C., were jolted awake in their seats when Dr. Eugene Peniston, a distinguished, soft-spoken clinical psychologist and one of the few African-Americans in the field of biofeedback, gave a talk on an unusual research project. In a cadence more like a Baptist preacher's than a research scientist's, Peniston described research that he and Paul J. Kulkosky, from the University of Southern Colorado, had conducted on alcoholics at the Fort Lyon, Colorado, Veterans Hospital, using techniques they had learned at the Menninger Clinic, the bastion of early deep-states biofeedback. The protocol guided users into a deep, trancelike state where alpha waves and theta waves, which signify deep relaxation, are elevated. Peniston found that those frequencies promoted a healing state that could be used to treat many things, from alcoholism to drug addiction to depression.

The controlled, randomized study used thirty men. Twenty were chronic alcoholics who were drinking themselves to death, all veterans and all back in the hospital for one more in-patient treatment regimen for alcoholism. These patients were divided into two groups. Ten were given conventional talk therapy treatment and took part in the twelve-step process, while ten others were given all of the traditional therapies, as well as a series of novel treatments that has since become known as the Peniston protocol. A third control group of ten nonalcoholics was also included. They received no treatment but had all the measurements taken.

First, the experimental group was given eight relaxation training sessions with standard hand-warming biofeedback. Members of the group were instructed in relaxation techniques until they could raise their fingertip temperature—cold hands are a symptom of anxiety—to 95 degrees in two consecutive sessions. Then they were given something called alpha-theta brain wave neurofeedback therapy. As is the case with beta and SMR training, alpha-theta uses a computer or analog instrument to guide users to a specific frequency or range of frequencies and hold them in that range. Yet alpha-theta is a very different approach. While beta and SMR can be thought of as more of a physiological approach, strengthening the cortex to alleviate symptoms and not concerned with their cause, alpha-theta training derives from the psychotherapeutic model. Lying back in a recliner with eyes closed, clients allow a therapist's voice and sounds from a neurofeedback instrument to guide them into a deeply relaxed "alpha-theta crossover" state in the alpha (8–12 hertz) and the theta (4–8 hertz) range, right on the edge of sleep. The active ingredient of the training, many believe, is the imagery of painful, repressed memories that the client has suffered with for years, which surface and are then resolved in this state. The name comes from the behavior of the brain waves being shown on the screen. Alpha waves are usually higher in amplitude, or more powerful, than theta. When, during this therapy, the amplitude of the alpha waves drops and the theta amplitude rises to the point where it crosses over the alpha waves—which means it has become more powerful—it's called alpha-theta crossover. It is a very specific state that is associated with the resolution of traumatic memories.

Before the twenty-eight-day course of treatment, all three groups were given an EEG and several standard psychological tests, including the Beck Depression Inventory, the Minnesota Multiphasic Personality Inventory II, the Millon Clinical Multiaxial Inventory, and the 16 Personality Factor Scale. Each group also had blood drawn to measure serum beta-endorphin levels. These hormones show marked elevation during times of physical and emotional stress. After a month-long treatment regimen, the three groups were given the tests again.

The results in the subjects receiving brain wave training were dramatic. Their EEGs showed a substantial increase in alpha and theta brain waves, a sign that they were more relaxed. Scores on the Beck inventory showed a sharp reduction in depression. The MMPI, the MCMI, and the personality scale all changed markedly in a desirable direction. On the other hand, the standard therapy group made only a few very small gains. The measurement of beta-endorphin levels in the blood illustrated a different kind of change. The group that had received traditional therapy experienced an increase in the beta-endorphins, which the researchers surmised was the result of stress caused by remaining abstinent from alcohol, the so-called white-knuckle approach to remaining sober. In the brain wave–training crowd, meanwhile, endorphin levels stabilized. The most dramatic difference, however, was that eight of the ten alcoholics receiving the Peniston protocol stopped drinking, while all ten members of the group that received traditional treatment were rehospitalized within eighteen months. Moreover, the abstinence lasted—three years later, only one of the eight had relapsed. In the field of addictions, numbers like this are unheard of. Twenty to thirty percent is considered excellent, and that number usually drops after six months as the temptation to use again becomes too much.

As an endorsement of the study, a paper on the Peniston-Kulkosky study, called "Alpha Theta Brain Wave Training and Beta Endorphin

Levels," was published in *Alcoholism: Clinical and Experimental Research,* one of the field's top refereed journals.

The study sent a tremor through the small world of biofeedback. Many welcomed the report as a new and powerful method of treating alcoholism and a tremendous boost to the field. It reaffirmed the power of deep-states work, encouraging a return to the alpha and theta training that had been popular in the 1960s and 1970s. Others, however, greeted it with a great deal of skepticism. The results were far too good to be believed.

And it opened old wounds. Deep-states training had swept the field of biofeedback before and had, rightly or wrongly, led to its downfall. "When the house of alpha collapsed," says Siegfried Othmer, "the biofeedback community picked themselves up, turned their back on it, and said they wanted nothing more to do with brain wave training. Now it was déjà vu." Although there are people who work with both alpha-theta and the higher ranges, beta and SMR researchers tend to be fairly straitlaced scientifically. Those who work in alpha or alpha-theta, on the other hand, tend to be much more liberal in their approach and beliefs, though they must do careful studies and interpret their work into statistical measures—the language of science—to have credibility, as Peniston did.

Alpha-theta brain wave training also works in strange and mysterious ways, ways that antagonize hard-core researchers. The language of the unconscious is metaphor and symbol, and people who see an iceberg or a flock of birds in their session will be asked to interpret what they see and how they feel about it. Reports from alpha-theta clients frequently go far to the edge of strangeness, and spiritual experiences—encounters with Jesus or dead relatives or other spirits—are not uncommon. People often have sudden bursts of insight and creativity. These experiences are not, clinicians say, to be dismissed as trivial products of the imagination or flights of fancy—rather, they are often deep psychological experiences, and having one can be powerful and transformative. "I've seen people in severe depressions simply have an experience of children playing [in the alpha-theta state] and come completely out of their depression," says William C. Scott, who taught the alpha-theta protocol to clinicians for EEG Spectrum and now his own practice. "If a relative died and there's some unfinished business between the two, about five to ten percent of people who do alpha-theta will have an experience of the bright lights, the tunnel, being there in the company of the loved one, the whole bit. Often there's a nonverbal exchange. The client will be there with guilt or shame or whatever, and there will be a nonverbal exchange, and the deceased will say something like, 'Don't sweat it, man. It's a game down there. You guys take things way too seriously.' And there is clearly a resolution. It's always cathartic." A catharsis, says Scott, that lasts.

It is also routine, experts say, for clients to become more intuitive in the state and "pick up" things the therapist is thinking or seeing. One therapist hooked up a client and retired to her office to peruse seed catalogs to plan her garden. At the end of the session when she came in to unhook her client, the woman said that her session had been a good one, but for some reason she had seen many different kinds of vegetables in her mind's eye during it. Scott recommends that therapists do twenty or more sessions of their own to process out their own anxieties and fears to avoid contaminating their client's session.

Peniston, motivated by horrific rates of alcoholism on the Ute reservation in Utah, where he worked in the 1970s for the Indian Health Service, continued his work after the 1990 conference, and alpha-theta again demonstrated remarkable promise. In 1991, he published the results of another study in a top journal, *Medical Psychotherapy*, using the same approach, this time with Vietnam veterans who had suffered combat-related post-traumatic stress disorder, or PTSD, twelve to fifteen years earlier. PTSD is a debilitating affliction caused by terrifying memories that often cause ongoing trauma such as nightmares, panic attacks, anxieties, depression, phobias, and flashbacks. A veteran suffering PTSD, for example, might wake up screaming every night with nightmares, wildly pulling his trigger finger, speaking Vietnamese, sweating and shaking profusely, believing he is in the midst of combat. Even though patients have these symptoms, they do not always remember the incident that caused them. It is estimated that PTSD affects nearly half a million Vietnam veterans.

There were two groups in this study: one, with fourteen subjects, received traditional therapies, such as group therapy, individual therapy, and psychotropic medications. The fifteen subjects in the other group received brain wave training in addition to the traditional therapies. By the end of the nearly monthlong study, only the experimental group tested within normal limits on the Minnesota Multiphasic Personality Inventory. Members' nightmares and flashbacks were significantly reduced, as was their psychotropic medication. Changes in the control group were far less dramatic and did not show a reduction in nightmares and flashbacks or medication. Thirty months later, according to a follow-up study, twelve of the fifteen veterans who had done the alpha-theta training were living a normal life, while all fourteen in the control group had relapsed.

Moreover, according to the study, the brain wave training allowed the repressed events that had caused the PTSD to surface to consciousness and be processed out, and to end the grip they had had on the patients. One patient had recurrent fears of being mutilated. While undergoing brain wave training he recalled the responsible incident, which had been completely repressed. While working with Security Forces during the war, he and a buddy were interrogating two captured Vietcong. The two enemy soldiers somehow killed his partner; the patient grabbed a shotgun and blew one Vietcong soldier's head off and beat the other man to death and dismembered his body. In another case, a soldier had recurring flashbacks at work, in which he would fly into a rage and have to be sedated. During alpha-theta he remembered that a friend had been wounded and the patient hid him in the brush, intending to return soon with a helicopter crew.

Darkness was closing in, and the crew said that they were fearful about returning and that the friend would be all right if he was left in hiding until morning. When the patient and the helicopter crew returned the following day, they found the friend's nude mutilated body hung by the feet from a tree. The patient had profound feelings of guilt for not retrieving his friend sooner.

The strength of the Peniston protocol, if it works as its practitioners believe it does, seems to be that it gets at the root of the reason people drink or take drugs or have emotional disorders. The protocol is in part a kind of computer-assisted psychotherapy, similar to, though ostensibly much more powerful than, talk therapy. Traditional talk therapy tries to get clients into a deeply relaxed state where they have access to unconscious material, which allows traumatic events to emerge and be processed. The alpha-theta protocol apparently takes the client to a very specific frequency portal—4 to 12 hertz—through which repressed childhood memories and other emotional painful events are easily accessed. A key difference between talk therapies and the alpha-theta protocol is that in alpha-theta the subject does not have to physically relive the trauma to exorcise it. Because the client's physiology is extremely quiet during an alpha-theta session, painful memories bubble gently to the surface; rather than reexperiencing it, clients in the alpha-theta state feel as if they are watching the trauma play out in front of them as on a television or movie screen, which is called the "witness state." A reexperience in such a witness state allows the event to become part of one's historical, narrative memory, rather than remaining emotionally reactive in the present.

No one knows what the physiological mechanism of the protocol is, but repressed memories and unresolved traumas apparently exert a stress on the brain that interferes with normal operation. Research shows that alcoholics, for example, cannot produce the appropriate type of alpha waves—the frequency that is characteristic of feeling relaxed and comfortable, sometimes referred to as an "internal anesthetic." Without the right kind of alpha, a person feels raw and exposed and edgy. Alcohol is a chemical that temporarily creates an artificial flow of good alpha—though it comes, obviously, with a raft of other problems. Somehow, the model holds, traumatic events block an alcoholic's brain from generating enough alpha. People who have gone through alpha-theta training say they feel more relaxed and may no longer have the overpowering urge to drink.

Bill Scott knows what it's like to crave alcohol and the alpha it creates. He was the second youngest of six children in what he describes as a dysfunctional working-class family in Hibbing, Minnesota. By the age of fourteen he was an alcoholic. For the next six years he tried to sober up many times, always unsuccessfully. Finally, he was given an ultimatum by a girlfriend: Sober up or it's over. A counselor he went to see convinced him to get into a thirty-day inpatient treatment program to dry out. A couple of other things turned him around. One was a realization in the form of a question: "Was I going to live life as a sick drunk or going to go into the unknown world of responsibility and use my power in a constructive way?" The second was the death of his older brother Mike, in an automobile accident, while driving home drunk from a party. "That really fueled me to want to do research," says Scott. At twenty he sobered up. He attended school at Bemidji State University in Minnesota, and after he graduated with a BSW he took the first of a series of jobs treating addictions. In 1991, he went to work directing a small chemical dependency program at Nett Lake, a tiny town on the Bois Forte Indian Reservation in far northern Minnesota. As on many Native American reservations, alcoholism was rampant, as high as 60 percent, and the program was in a state of near collapse. In two years there, Scott referred 160 people to treatment facilities for alcohol and drug problems. "Out of 160, I would say two people had significant sobriety," he says. With such a dismal success rate, Scott says, "I was constantly on the lookout for alternative treatments." In 1993, Scott read Peniston's first paper and was intrigued. By coincidence, on the same day a colleague called and asked him if he had ever heard of the Peniston protocol. Scott laughed, considered it a sign, and they decided to refer to a clinic that offered Peniston's approach in nearby Grand Rapids, called Northland Mental Health. The first referral was a long-term and severe alcoholic, a Native American man, who had been through several treatments. "I went to see him two weeks after he started," recalls Scott, "and he was very different—calm and much more centered. I could really see the changes in him. I was really amazed." "It's too good to be true," Scott thought, and to test the process further he referred two more chronic alcoholics. "They had some of the most severe personality disorders I have seen to this day. My attitude was 'This can't really work. Let's see what they can do with these two.'" To his amazement, he says, "The game playing stopped within a week. They became more open and more honest, and one guy told me, 'I started noticing people. I started noticing everything. I was more present.'"

Extremely intrigued, Scott read up on the technique, devouring the limited literature available. Then he accepted a job directing the Northland alpha-theta program and got eighty hours of instruction from Dr. John Nash, the director of the program, who now directs a clinic in Edina, Minnesota. Scott also did twenty hours of his own alpha-theta training, which he says had a powerful effect. He had a panic disorder, and when it appeared—almost every day—he couldn't function: his heart pounded, he felt like the walls were closing in, he obsessed on things, couldn't eat, and would have to go to the nearest bathroom to pull things together. "Just two sessions took a bite out of the panic attacks," he says. By the time twenty sessions were over, the panic disorder had vanished.

In late 1993 Scott phoned Peniston (who by now had moved to a VA center in Bonham, Texas, where he was not allowed, for bureaucratic reasons, to practice his protocol), introduced himself, and asked Peniston if he would supervise him over the phone in an uncontrolled study Scott would conduct at Northland. Together they wrote a design. It was the same protocol that Peniston had conducted:

Reward 8–12 hertz alpha and 4–7 hertz theta. Later they added the inhibition of 2–5 hertz, or delta, to keep people from falling asleep. There were twenty-four alcoholics in the study. After a month of treatment, fourteen of the twenty-four stopped drinking, Scott says. "Ten attempted to use. Five attempted to use very briefly and got sick and got back into recovery again." After twenty months, Scott left the field and lost track of his clients. But nearly two years after the study ended, he said, the results were close to what Peniston had achieved.

Burned out by his work in the field of chemical dependency, in 1995 Scott went to Los Angeles to take a job in the field of computerized graphic design. Then he heard about EEG Spectrum. The Othmers had been trying to incorporate alpha-theta into their practice, but they had some problems fully implementing it, and some therapists were apprehensive about it. "There's a huge therapist influence in this," says Scott. "If the therapist has a lot of anxieties and fears, the clients pick those up when they go into those vulnerable states. Then they stop going into those states." Othmer hired Scott to run the alpha-theta program.

One of the first things Scott did when he arrived at EEG Spectrum was look for a big treatment facility to do a large-scale, long-term study on the effectiveness of alpha-theta training, not for alcoholics, but for serious, strung-out drug addicts. It is a population known as poly-substance abusers—they have a primary drug of choice but also use other drugs, especially alcohol and marijuana. Addicts pose a different problem than alcoholics. They have too much of the wrong kind of alpha present in their EEGs, rather than too little alpha, which tends to make them not present, spacey, and removed from the world around them. The challenge is to lower their "bad" alpha and bring up the "good," though the protocol is similar.

Scott talked to people at an addictions treatment center called CRI-Help, and they were interested. The chairman of the center's board, Marcus Sola, who is also the vice president of a mental health HMO in Los Angeles called the Holman Group, had done some beta/SMR training at Spectrum with Sue Othmer. "I had an immediate result response for some anxiety I had," says Sola, "and I immediately purchased a system, and we began offering training to the staff at Holman." When Scott called about a study at CRI-Help, he already knew of the power of neurofeedback, he says, and convinced the board of directors to sign on.

CRI-Help started in 1969 as a group of addicts who were determined to quit, and who lived together in a house to provide mutual support for one another to work through their withdrawal. "The only rule," says Scott, "was don't get loaded." The place thrived, and now the facility is a 120-bed, state-of-the art, residential treatment facility in a brand-new, 35,000-square-foot, stuccoed building in North Hollywood that treats every kind of chemical dependency, from heroin to cocaine to alcohol, primarily on the principles of the twelve-step program of Alcoholics Anonymous and Narcotics Anonymous.

CRI-Help is anything but an old-style place to go cold turkey. It is a comfortable and supportive place to heal the human mind, on both an inpatient and an outpatient basis. Each therapist at the center has no more than ten clients, a remarkably low client-to-therapist ratio. The rooms are simple and clean with plenty of natural light and either two or three beds to a room. The lobby has a spiral staircase and sculptures, and a mammoth built-in aquarium on one wall. Upstairs is a large, completely empty, round room dedicated solely to meditation. There is a place outdoors for sports, including a basketball court, as well as a barbecue and tree-shaded picnic area, where, on the day I visited, birds trilled in the sunshine.

Scott and staff therapists tried the alpha-theta technique informally on six patients. "It became quite apparent that the same kinds of things were happening that had happened in Minnesota," he says. "We saw the same kind of changes." After several months, Sola was sold on the proposed study, and Scott, with some help from David Kaiser, a Ph.D. cognitive neuroscientist at EEG Spectrum, and with the endorsement of the Human Subjects Review Committee at UCLA, which must approve the ethics of the study design, wrote two grant proposals to the National Institutes of Health for $400,000 for two years. Both grants were rejected, and CRI-Help ended up funding the study on their own, with some help from EEG Spectrum, for a total of around $320,000.

The study had fifty-five people in an experimental group and fifty-five in the control group, a mixture of men and women, selected at random. The study was about a third heroin users, a third crack cocaine, a third methamphetamine. Each subject had been free of his or her drug of choice for fourteen days at the beginning of the study. Before they started any treatment, all subjects were given standard psychological measures, including a Test of Variable Attention, which measures how people pay attention and respond. Treatment with alpha-theta, as with all therapy, is an art as well as a science, and Scott's approach varies slightly from Peniston's. Scott's work weds the SMR/beta training that EEG Spectrum uses with the alpha-theta; he believes that part of what helps prevent retraumatization as buried memories emerge is at least ten sessions of beta or SMR training to stabilize the nervous system, an element that is not part of the Peniston protocol. Because of the use of the high-range training, the CRI-Help protocol did not include hand warming or deep breathing to relax the subjects before their sessions.

In November of 1996 the study got under way. After the higher-range work was done, clients were taken to a small, quiet room, where they lay back in a comfortable reclining chair. The frequency the client is trying to access is 4 to 12 hertz, which includes the theta, or hypnogogic, state, the window through which painful and repressed childhood memories emerge. Many who have had traumatic memories cannot at first easily access those frequencies. They close their eyes and their brains respond automatically, and eventually they learn to relax. With Neurocybernetics equipment and software, as a person's EEG floats to a frequency within the 5-to-8-hertz range, the theta, they are rewarded with a pleasant bonging sound. When they get into the

8-to-12-hertz alpha state, they get a pleasant binging sound. The idea is to float back and forth between the two states. "I tell them it's as if they are walking down a road and there is fruit on both sides. They want some of each," Scott says. The brain likes the pleasant sound of the bings and bongs, and when they are not happening, it learns to shift back to the state where it is getting rewarded. The process discourages, or inhibits, sleep and muscle tension by turning off the pleasant sounds if the client stays too long in either of those states. Additionally, the pleasant sound of crashing surf occurs when the person is transitioning to theta, and a babbling brook occurs when the person is headed into alpha. Everyone passes through these frequency ranges every night on their way to sleep; the equipment merely helps clients to dwell there. "People who drink or use drugs are in something akin to a trance," Scott says. "This teaches their brain how to come out of it." Experimental subjects underwent two forty-five-minute sessions of alpha-theta neurofeedback a day, for three weeks (ten to twenty beta/SMR and thirty alpha-theta).

Two things are going on during a session. First, as the physiology quiets, resistance to allowing the memories to surface declines. "They go deeper and deeper in each session," says Scott, "getting closer to the memories, closer to what's going on." Scott says the mind has an innate ability to release only the amount of information that can be handled and doesn't pour everything out all at once and overwhelm the person. "And because they are in a quiet state, people don't have a flashback or reexperience the trauma," he says. "They become emotional, and they'll cry, but they won't reexperience. They process it cortically." Once the memory is reprocessed cortically, in the conscious mind, the shock to the brain seems to be largely over. Scott says that there is no need for a client to have imagery to benefit from the therapy, and that profound alpha-theta experiences happen in only about 40 percent of the clients. "People who have PTSD and do sessions, whether they have imagery or not," he says, "no longer meet the criteria for PTSD."

The second thing that Scott and Peniston claim is happening during these sessions is that a person's subconscious mind is being rescripted. Thomas Budzynski's and Johann Stoyva's research for the military in the 1960s showed that people in a theta state were "hypersuggestible," as people are in a hypnotic induction, and learned new information extremely rapidly and uncritically, apparently because being near a state of sleep enabled patients to bypass the conscious mind. The new "script" of ideas on how to avoid drugs or alcohol that is given to clients just before they go into alpha-theta is still fresh in their minds, and when they get to the deep state, it very rapidly becomes part of their belief systems. The new script becomes so ingrained, in fact, that about 25 percent of the people who take drugs and half of the alcoholics who undergo alpha-theta training and try to use again get physically sick when they do, an illness that's been dubbed the Peniston flu. "And if you have ever seen that with a patient," says Peniston, "boy, you know they're sick. It don't take much." Others report that they no longer get the high they once got from their substance of choice. A neurofeedback therapist in Oslo, Oystein Larsen, treated a woman who was a heroin addict. Partway through the treatment she quit the program and went into town for heroin. The drug no longer gave her the accustomed high, he said, and she threatened the therapist with a lawsuit.

I did several alpha-theta sessions with Bill Scott at the Spectrum offices in Encino, and he treated me as if I had a chemical dependency problem accompanied by a panic disorder so I could experience the protocol as the people in the study did.

"Okay, Jim, sit down here in the chair and close your eyes," he said. Scott, thirty-three, has dark hair and a goatee and a gentle and reassuring air about him; charisma is an unmeasurable variable that is important to any psychotherapeutic protocol. This echoes Joe Kamiya's claim that a rapport between the subject and the researcher may be the most critical element for a subject to produce alpha. The subject has to feel relaxed, supported, and confident enough to let go and ease into the subtle, somewhat vulnerable state.

In a small, dark room, lit only by the blue glow of a computer screen, Scott covered me to the neck with a blanket, and placed an eyebag over my eyes. I faced a computer screen with speakers, though I couldn't see it. Scott hooked an electrode to what EEG technicians call the Pz site, in the center of the back of my head, just above where it hits the pillow. The biofeedback started, and Scott began with the visualization that, coupled with the alpha-theta brain wave training, he believes is the most active ingredient of the Peniston protocol. It is very similar to scripts used by hypnotherapists. "Picture in your mind your ideal life," he told me, sotto voce. "What characteristics would you like to further enhance? More inner peace, more calm when you drive. You want to be free from your addiction to chemicals, happier and more content. You want to reestablish a relationship with peers and your family." Scott paused for a moment to let that sink in. "Now picture what your life would be like if you were to resume your relationship with chemicals, knowing what has happened with previous treatment experiences. You mentioned you are on parole and you get urinalysis tests, or you get caught doing burglaries. Imagine what's likely to happen over the first month. The first three months. The first year. All the way to the end. And imagine what you think the afterlife might be like as well." He paused again to let me form that imagery.

Then Scott got into what is called an abstinent projection scene. "Imagine yourself in ongoing recovery. What a healthy life might be like. Imagine going to meetings on a regular basis, meeting with your sponsors, continuing to allow people to get to know you, your loneliness in groups beginning to dissipate, your loneliness level beginning to decline, relationships restructuring. Imagine what good things are likely to happen a month from now, three months, a year. And feel free to imagine your understanding of the afterlife." Another pause.

"Now imagine Uncle Timothy coming by, and he has your favorite chemical. [This part of the script was based on interviews with clients before the sessions, in which they describe how the pressure to use drugs again might surface.] He says, 'How's my little buddy! My little nephew!' Then he says, 'I got your favorite stuff here.' You think, 'I don't want to disappoint him,' and you think of the first time you used with him and how wonderful it was. Imagine yourself feeling the discomfort in that situation but really taking care of yourself in that scene. Imagine yourself saying to Uncle Tim, 'Nah, I can't handle this stuff anymore. I'm not doing that. I'm not going to use.' Imagine him saying, 'What? Are you going to those little AA meetings again?' And imagine yourself saying, 'Yeah, I'm starting to take care of myself.' Imagine yourself having to leave or go for a drive or letting him know it's very difficult and asking him to take it away. Imagine yourself resolving it in a way that you won't have to use." Scott paused again, this time to shift and deal with the panic attack part of the pathology. "Then I ask you to imagine yourself in the car, driving to the store to pick something up. Imagine yourself getting into the car and being calm and relaxed. As absurd as it might be, envision it. If you can't imagine it, think the words 'I am getting into my car and driving to the store to get some iced tea. I thought when I got to Balboa, I might have a panic attack. But I passed the Balboa exit. I felt myself tense a little bit, but I remain calm. I get to the store, buy my iced tea, and drive home and find myself back without any problems.'

"Imagine that all of these scenes that we've covered, that they are already stored on special videocassette tape, one that your subconscious mind knows how to play and replay. Imagine that you're holding this tape as a whole in your hand. Imagine yourself bringing this tape up, touching it on your head, and as you feel the pressure on your head, imagine the tape turning into light, traveling through your skull, going into your subconscious mind, all as a whole, communicating to the subconscious your intentions to give you the experiences you need to resolve any unconscious conflicts you have, that are barriers." He paused. "Now picture yourself in a safe place you've created, a

Hawaiian island. You're feeling very secure, very safe, very happy. Content. Very satisfied. Allow that scene to remain in your mind for a minute or so." Now Scott prepared to begin the feedback session. "Allow your mind to clear now for a half hour or so. In twenty-seven minutes from now I'll let you know you have three minutes left, and after that I will invite you back. Until then, just enjoy."

Scott will sit in the room with the client if he is treating only one client at a time. "If someone has issues of abandonment, you have to stay," he says. "Others don't want you there. Women who have been abused, for example. They are not comfortable going into a vulnerable place."

For half an hour my mind wandered in a dreamy twilight, in and out of sleep, barely aware of the pleasant sounds. Toward the end of the session, Scott returned and said softly, "Three more minutes. Slowly start bringing yourself back. Whenever you're ready, you can start sharing your experiences." Scott says he encourages people to keep their eyes closed. "I never ask them to describe their imagery. Then they exclude everything else. Sometimes people will have a *feeling* of driving down the road and a *feeling* of a hole. Or a *feeling* of floating. Or they can recall dialogue between two people."

It's hard to describe this twilight state because by nature of the experience one's conscious mind is disengaged and the world is an internal one. But there seemed to be a free-form cascade of imagery of every type, mostly bits and pieces, though occasionally a rather strong memorable image drifted through. I told Scott that I had a strong image of myself driving down a road, when a hole appeared ahead, and I was fearful of hitting this hole because I might lose control. Processing the experience is critical, Scott says, and in my case he bounced my imagery back to me in a way that would get me to think about what I had experienced. "Do you have the scene in mind? Can you recall any of the emotions? What is a hole, and what do you think that represents? What is your home, and what does that represent? What is a car? What is control? How will you stay on the road?"

As I answered those questions, Scott did not try to get me to apply these things to my specific circumstances but simply to create a critical mass of thought about them, and to let them do their own work in my mind. "The issues often just come up," he said. "People will spontaneously bring up the issue the imagery is about."

The most essential aspects of the process, Scott says, include the five phases of imagery: constructing the ideal personality, envisioning recovery, rejecting the chemical, instruction to the unconscious about handling the material, and imagery of safety and well-being. Simply "stewing" in Alpha-theta is not, by itself, enough to treat serious addictions, experts say. Also vital to success is the twelve-step process, which usually involves a sponsor who knows the addiction and can counsel the addicted, as well as regular group contact. But it is key, says Scott, that the brain has been taught, through operant conditioning, to produce more alpha, to feel more comfortable. That is why the craving usually disappears and why the therapy is effective. "People who do neurofeedback stay in treatment twice as long as people who don't," says Don Theodore, a longtime counselor at CRI-Help. "And the longer you keep them in treatment, the better their chances."

The results of the CRI-Help study were dramatic, Scott says, nearly as dramatic as the work of Peniston and others. The study hasn't been published yet and exact numbers have not been released. For people who were out of the treatment program from three to twenty-three months, Scott says, the experimental group had a recovery rate four times that of the control group. Experimental subjects also stayed in treatment longer. The paper was published in the *American Journal for Drug and Alcohol Abuse* in 2005.

"It's quite impressive," says Sola, of the results of the study. "It's clear that it has increased the success rate in outcomes and length of stay. It's an incredible enhancement to treatment."

What is happening at a cellular level with the alpha-theta protocol? Bill Scott's model, based on research by Joseph LeDoux at NYU, is applicable. When someone is faced with a trauma—say a bomb explodes in a crowded restaurant—there isn't time for the information to come into the brain, travel to the cortex, and be sorted out. Instead, an innate survival mechanism takes over. The cortex is taken out of the loop, the information goes directly to the amygdala, which governs fear response, and the person automatically flees the restaurant. Even after the trauma is over, the amygdala will sometimes continue to be engaged directly and create inappropriate fear responses. (Scott believes that this fear response also takes over when addicts enter treatment because they are about to be deprived of the substance on which they're dependent.)

It is conjecture, but Scott thinks that as the alpha-theta training helps release emotionally traumatic memories, which in turn eliminates noxious stress chemicals associated with the trauma from the brain, cellular function in the cortex is allowed to normalize. Once that happens, the cortex comes back on-line and re-engages in the processing loop. No longer are memories being made from a predominately fearful and anxious state.

There are a couple of interesting footnotes to the study. Scott claims that the EEG is a kind of lie detector: As the treatment progresses, if a person is getting well, his or her alpha and theta waves grow closer together in amplitude. If a person is heading toward a relapse, the alpha and theta rhythms grow farther apart as they access the old "procedural" memories. Once that warning begins to show up, Scott says, that person can receive some type of intervention to help him or her stay in treatment. "Alpha-theta also changes the way people dream," says Scott. "Almost everyone remembers their dreams after alpha-theta." Clients also often report a greatly enhanced creativity.

Another large-scale study that may soon be published was done by those who carried on the legacy of Elmer and Alyce Green at the Menninger Institute. Steve Fahrion, Ph.D., and his wife, Patricia Norris, a Ph.D. and past president of the AAPB (who is also Elmer and Alyce Green's daughter), left Menninger in 1994 when, because of budget problems, the institution cut a number of things, including brain wave training. Fahrion, Norris, and others formed the Life Sciences Institute of Mind Body Health, both to treat clients and to do research. The most recent research was a study with 810 convicts in the Kansas State Prison using the alpha-theta protocol. All of the group received traditional treatment for alcoholism. One third also received the Peniston protocol, including the alpha-theta brain wave training. After the prisoners were released, they had to maintain sobriety, and checks included random urinalysis tests. "There is a significant difference between the groups that favors those who received alpha-theta training," says Fahrion. "The results were three times as good as those in the traditional treatment." As was the case with Scott, Fahrion has found that people with depression routinely normalize with alpha-theta training.

Roger Werholtz is a deputy secretary of corrections in Topeka who oversees medical and psychological programs. The Department of Corrections recently expanded the brain wave program to two other prisons, and he says it has promise. "We've seen enough positive evidence to convince us to take more of a look at it," he says, "but not enough to convince us to make wholesale use of it." Three years after treatment, in a population of 109 prisoners, the rate of relapse was about 53 percent for the brain wave group, and 67 percent for the traditional therapies. The most impressive aspect of the training was that many more people in the brain wave group finished therapy, which is a common report. "It's expensive," Werholtz says, "but measured by the cost of successful completions, it's the cheapest program." Cognitive therapy has been shown to be effective in a prison setting, and Werholtz says that the prison system is now comparing the effectiveness of cognitive therapy to alpha-theta work.

Two things impressed Werholtz about the Peniston protocol, he says. One was a personal experience during a weekend workshop at Menninger. "Enough happened out there, with them leading us into altered states and our own experience, that it convinced me it was a powerful intervention and we should try it." Second were reports that alcoholics and drug addicts who went through the program became physiologically sick when they tried to use. That did not happen in Kansas, Werholtz says. But he still believes it is possible, and so fine tuning of the methods continues.

Many experts who actually work directly with addicts are also more hopeful. For Bob Dickson the Peniston protocol is a godsend. A master's-level counselor, from 1987 to 1993 he served as the director of the Texas Commission on Alcohol and Drug Abuse. He witnessed the profound changes the Peniston protocol brought to violent offenders in the Texas penal system when he commissioned three separate studies using the protocol. Dickson took early retirement in 1993 and went into private practice doing EEG neurofeedback; he is now working with the Southwest Health Technology Foundation to replicate Peniston's work, not with alcoholics but with homeless crack addicts. "Nothing else in the field of addictions works as well," he says. "Our mission is to replicate the work and get it accepted by the insurance companies and private corporations."

One of the few psychiatrists in private practice using neurofeedback, including the Peniston protocol, is Dr. J. Alan Cook, in Mount Vernon, Washington. A board-certified psychiatrist for the past twenty-eight years, Cook has used neurofeedback training in his practice for the last two years and has treated some forty people. He uses both the higher-level and deep-states training, and both, he says, have become fundamental to his practice. "It's one of the most important things that I do," he says, along with cognitive therapy and medication. "I'm impressed with the results from neurofeedback." Cook hasn't treated many alcoholics or addicts; he finds that difficult in an outpatient setting. But the results of combining beta and SMR with deep states for depression and other problems have been remarkable. One woman had been coming to see him for seven years, for an hour once a week. Talk and medication helped her, he says, "but she never did quite break out of the things that held her in bondage." After a course of deep-states training, "she has broken through that," he says. "The clinical improvement is just miraculous." Cook also had one client, however, who, despite dozens of sessions of both kinds of brain wave training, failed to see any change at all in his symptoms.

Criticism of the Peniston protocol comes from a number of different angles. Dr. Norman Hoffmann is a Ph.D. psychologist at Abt Associates in Cambridge, Massachusetts. For the past twenty years he has evaluated studies of addiction treatments and has gathered a database on more than eight hundred scientific studies. When we talked on the telephone, he told me that he had never heard of the Peniston protocol, but he responded to a description of Peniston's work. A major problem, he said, is the number of subjects. "I tend not to put credence in anything less than 100 people," he said. "Preferably, the study would have 250 to 450." A thirty-person study is "a nice pilot finding that would need to be replicated for two reasons," he said. "You want to have a bunch more folks—300 at least. Second, you'd want to have it done by someone who is trained by Peniston but not him. Some people in psychotherapy know the art well. If we can't make the art into a science, it has limited utility. All we can do is go out and find those people. We can't teach others to do it."

There has also been criticism from inside the field of self-regulation, even by scientists who believe in the power of neurofeedback and who would like to believe that the Peniston protocol is an effective way to dispense with painful memories and cure addictions. But the studies, they say, have fundamental flaws. Dr. Edward Taub, a professor of psychology at the University of Alabama at Birmingham, is one of those critics. He did a randomized, double-blind, controlled study of his own with biofeedback at the Rehabilitation Center for Alcoholics and District of Columbia Veteran's Home, in the 1970s, though it was not published until 1994. The study had four groups. Each received routine therapy, and in addition one group did transcendental meditation; one did EMG, or muscle biofeedback; and one received treatment with something called a Russian sleep machine, which has electrodes that attach to the forehead and creates a slight feeling of current passsing into the head (though Taub says it doesn't induce sleep). One group received only the routine therapy. A year and a half later, the routine therapy group had 33 percent complete sobriety, while the EMG and TM groups had 66 percent—close to Peniston's numbers.

Taub has several criticisms of Peniston's approach. First, the studies utilize a "treatment package." "It's not just alpha-theta biofeedback," he says. "There's thermal biofeedback, guided imagery, cognitive therapy, and I think there's even progressive muscle relaxation. The modality that is focused on is alpha-theta biofeedback. I'm not calling into question if the Peniston protocol is effective—I think it is. But we don't know which component or components of it are effective, and there's no evidence whatsoever that it's the alpha-theta feedback that is prepotent." Peniston's studies are also missing at least two critical elements, he says. First, there is no control group that received some type of high-tech placebo control, something that would make the subjects think they were hooked up to a sophisticated machine that was going to make them sober, such as the Russian sleep machine. Second, he says, "Peniston is a very charismatic individual, and there is a huge demand characteristic that this is going to work, and that is communicated to the patient." The control group did not receive the same kind of attention from Peniston or another charismatic individual, he says, which could also affect placebo outcome.

Another critic of the protocol is Ken Graap, a Ph.D. candidate in the Psychology Department at Emory University in Atlanta. A few years ago he and some colleagues there tried to replicate Peniston's success with PTSD. Graap admits there were some major differences in design: the patients were outpatients; they were older than Peniston's; they continued to take medication; and half of the twenty-one subjects left before the study was completed, which may have skewed the statistics because the sample size was too small. But the treatment included the alpha-theta protocol. "We didn't find any changes in

MCMI or MMPI," says Graap, which is where Peniston found dramatic changes. And they found that the control group members who were waiting for treatment had roughly the same response as the experimental group. "The data we got did not show an effect for treatment per se." Graap says that he was deeply disappointed, for he spent a great deal of time on the study and believed in the treatment, but he maintains that "no one study can make or break this protocol."

Something else has troubled Graap and his colleague, Dr. David Freides, a psychologist at Emory, about Peniston's work. They went over the Peniston and Kulkosky papers carefully, they say, and discovered a number of apparent errors and coincidences that raise questions that have not been explained, which they gathered and wrote up in a paper of their own. The study criteria required that participants not take psychotropic medication, for example, yet the authors reported that four veterans in the study "were withdrawn from tricyclic anti-depressants during the study," which indicates they were on medication. In another place Peniston reported that "Omni-prep" was used as a conduction paste on the end of the electrodes that pick up brain waves off the skull. The problem, the authors wrote, is that Omni-prep is an abrasive scrub, not a conductive medium.

Though these are trivial matters, probably oversights, Graap believes the field of neurofeedback will never be accepted by mainstream science until problems such as these are addressed. "I think the treatment overall makes good sense in terms of behavioral medicine," says Graap, "but until the data are cleaned up and the piers of support are there, in a very clear way, and you can use these concisely in a grant application, it's very difficult to convince funding agencies that they should go down this road."

Peniston, for his part, says Graap was not trained in the protocol. His failure to replicate was not surprising, given that he deviated from the methodology in several key ways, especially the fact that the subjects were still taking psychotropic medication. "To be effective,"

Peniston says, "the learning must take place in a drug-free state." Based on what Graap has reported, says Peniston, "I can assure you he did not employ the Peniston-Kulkosky protocol accurately."

While there are claims of replication of Peniston's work, including Scott's study, only case studies have been published. Peniston says that a large-scale, multisite, national study is vital "if we are to achieve the support necessary from the medical establishment and the health care delivery infrastructure, which will allow everyone who needs this assistance to obtain it affordably from properly trained clinicians."

Peniston's work is only one example of the therapeutic use of deep-states training, though it has gotten a great deal of attention because his studies are thorough, well designed, and have been published in prestigious scientific journals. Other work being done in deep states grew out of the original promise that biofeedback held in the 1960s and 1970s. Dr. Lester Fehmi and a handful of other long-time alpha researchers say that alpha brain wave training was very young and still evolving when it first appeared on the scene, and the hype got way out beyond the science.

According to Fehmi, when the backlash against the extravagant claims being made for biofeedback eroded scientific and financial support for the field, the baby was tossed out with the bathwater. He insists that the phenomena of deep relaxation—the release of stress and anxiety and resulting feelings of calmness, even transcendence—were very real. Now, Fehmi and others say, with the advent of computerized biofeedback in the 1980s and thirty years of evolution in the methods, alpha brain wave training is a very powerful and replicable phenomenon, a natural healing state that has come into its own. "We've been sitting on this diamond for thirty years and no way to get off of it and show the world, because no one believes it," says Fehmi. "But alpha training changes people's lives." (Alpha training is incorporated into the Peniston approach, which creates feelings of profound relaxation but usually doesn't include memories or emotional trauma common in the theta state.)

Fehmi's private clinic in Princeton, New Jersey, is adjacent to his modest suburban home on a busy two-lane, tree-shaded road. A Ph.D. psychologist, a former college professor with a physics and engineering background, and a veteran Zen meditator, Fehmi is an affable, soft-spoken man; his words are slow and measured. Bearded, wearing a jacket and a tie, he bears a resemblance to the motion picture actor Robert De Niro. He works with his wife, Susan Shor-Fehmi, who is a licensed and certified psychotherapist and is trained in Fehmi's brand of brain wave training as well. Their warren of office rooms is crammed with assorted electronic gear, tape recorders, compact disc players, and strobe lights, all attached with a rainbow of electronic wires. Fehmi is highly regarded in the brain wave training community and is one of the world's top experts on what is called synchronous alpha training, which he discovered in his days as a young assistant professor at the State University of New York, Stony Brook.

As a graduate student from 1961 to 1966, Fehmi studied visual information processing in monkeys at a laboratory at UCLA with Donald B. Lindsley, a pioneer in the study of the electroencephalogram. As the animals sat in chairs, a special projector flashed images of diamonds and squares onto backlit panels for just a few thousandths of a second. If a diamond appeared, the animal could press a button and get a banana pellet. If a square appeared on the panel, the animal could not get rewarded. The monkeys learned the task successfully 100 percent of the time.

The researchers knew that a flash of light presented right after the diamond or square could interfere with the neural information of the shape. But what was the minimum time needed between the information flash and the blanking flash for the light to have effect? They flashed the light 250 milliseconds, 200 milliseconds, then 100 and 50 milliseconds after the symbol. The monkeys still responded correctly. Only at 26 milliseconds did the monkeys start to guess, suggesting they weren't getting visual information. Researchers were startled, for it seemed far too little time for the neural activity to hit the optic nerve and pass through the circuitry of the brain. They were, however, assuming incorrectly that the information traveled serially, or from the eyes, down the optic nerve, to one site in the brain, to another, to the next. Instead, electrodes at different sites on the head that picked up the signal showed that the brain was processing the information simultaneously at all of the sites, something called synchrony. It was key to understanding how the brain—human as well as animal—is able to process information so quickly. A brain wave, like all electrical signals, is a wavy line, and synchrony means that the signals were reaching the top and bottom at the same time. Understanding synchrony, says Fehmi, turned out to be fundamental in understanding the human brain's electrical characteristics.

In the old days, alpha brain wave training was often done with two sensors on the brain, a bipolar placement. The values of the signals from the two sites were subtracted from each other, which, Fehmi says, was a complicated process and often led to mistakes in feedback and some of the poor research results. For a while Fehmi used a unipolar placement, which he says is simpler and leads to a better measure of synchrony. But his work with monkeys convinced him that the key to alpha production was in synchrony: training the whole brain at the same time so it was producing the same frequency and peaking at the same time. These synchronous brain waves, measured independently at five different sites on the head, have much more power than any other type. While beta waves are operating at higher frequencies in different parts of the brain at the same time, synchronus alpha waves are a smooth hum over all the brain. Adam Crane, founder of the Ossining, New York, company American Biotec, which has manufactured and sold a computerized synchronus alpha training system called Cap Scan since the 1970s describes alpha synchrony this way. "If children were swinging on a swing and they were holding hands, they're synchronous," he says. "If you get four or five swings in a row and the swinging is synchronous you can uproot the swing—because it is very powerful. Synchrony is related to an increase in power."

Fehmi got interested in biofeedback as a result of the alpha work of Joe Kamiya, whom he met at a cocktail party. In 1967, Fehmi left UCLA to become an assistant professor of psychology at the State University of New York at Stony Brook. With the help of technical staff people from the department of psychology he built some of his own EEG biofeedback devices and sought to wed the idea of synchronous brain wave activity to biofeedback. First, he hooked himself up to an EEG monitor and tried for two hours at a time to produce synchronous alpha. He tried self-hypnotism, relaxation, and meditation. But through twelve sessions, no matter what he tried, he was unable consistently to increase the production of alpha. During the thirteenth session, he recalls, he was so exasperated he thought to himself that the task was useless. "The minute I did that," he says, "a high-amplitude alpha brain wave pattern appeared on the EEG." Surrendering had brought the desired result. Success, he realized, had to do with how he paid attention; instead of concentrating so hard and focusing intently on his goal, he relaxed his focus and let go. This opening became a key to his therapy, which he calls Open Focus.

Profound feelings began to wash over Fehmi and his life, he says, began to change dramatically. He felt far more relaxed and at ease in his body, and, he says, "I had a feeling that I was walking on air." His "obsessive compulsive researcher professor" nature softened, and he adopted a more relaxed, more centered attitude toward life. Things that irritated him no longer even registered, and it enhanced his intellectual life. "I was teaching hard-core physiological psychology and research courses with an effortlessness that I had never before experienced," he says. "I laughed a lot more, and I was less serious." He had a newfound intimacy with his body and emotions, and he moved and played racketball, a favorite sport, with a grace and ease he did not have before. He also experienced a renewed connection with others—family, friends, and students all responded to these changes in a warm way. "I felt that I had a magnetic personality," he says. Physical changes happened as well. Severe, long-term arthritis in his hands disappeared, and his vision and sense of smell were enhanced. What was interesting, he says, was that he had not even been aware of the tension he had been carrying with him. It struck him that this was the way life was meant to be.

In 1973 Fehmi joined the clinical practice of a psychiatrist named Paul Webber in Princeton, New Jersey. For three years the two used brain wave training with patients to treat pain, anxiety, and stress. During the early days of his work as a researcher, Fehmi made another important discovery that would become part of his Open Focus approach. As part of the search for a way to help people generate alpha waves, he gave students a twenty-eight-question relaxation imagery inventory to determine if they produced EEG changes. There were two key questions on the inventory, which people answered while they were hooked to an EEG instrument that read their brain waves. The first was "Can you imagine the space between your eyes?" and the second was "Can you imagine the space between your ears?" All the other images—of golden sunsets or roses—were powerless in increasing synchronous alpha. Fehmi found that when people simply imagined space with their eyes closed, their production of synchronous alpha instantly increased—it was a powerful relaxation technique. He explained it like this: When someone is looking at an object, or even imagining an object, he or she is engaging many different parts of the brain to make sense of that object—via memories, senses, etc. In that state the brain is "more desynchronized"—that is, work is occurring at different frequencies in many different parts of the brain. As soon as the eyes are closed, one is imagining space, and there is a complete absence of images; the whole brain stops working hard and synchronous alpha takes over. It is a healing state, Fehmi says, and has a powerful effect on stress. The repression of physical and emotional stress relies, in part, on muscle tension, from the head to the feet—tension that usually has been present so long people don't realize it exists, in the same way people adapt to background sound and don't even notice it is there. No one knows precisely what is going on with alpha training, but Fehmi's model holds that synchronous alpha provides an absolute resting state for the muscles and, with time and practice, relieves stress and tension that they have carried for years. When this deeply relaxed state takes over the brain, human systems "let go" and reset. This is not unlike meditation, Fehmi says, only faster and easier for some.

Fehmi believes that a fundamental problem in our culture is that we are taught as small children to focus too narrowly on the external world. With our mind and eyes we grip objects and the outside world too intently. After a while the eyes, the brain, and other parts of the nervous system break down from being in such an intense mode. "If you gripped your hand as tight as you could all the time, after a while the muscles in your hand would freeze and not function," he says. "That's what happens with the rigid way we pay attention." And it leads to anxiety, depression, and a number of other problems. Fehmi coaches people in methods to relax their attention, and as they learn to do that, the whole physiology relaxes as well. The synchronous alpha brain wave training equipment is a powerful way to teach someone what it feels like to let go and how to maintain that feeling.

In Fehmi's Manhattan office, high above the bustle of Broadway, he hooked me up to his equipment. His present-day protocol uses a computerized biofeedback device with a strobe light and sound biofeedback. I sat in a chair and had saline sensors fitted to five positions on my scalp, with a headband to hold them in place. Eyes closed, I faced a strobe and could see it flicker through my lids. I donned earphones and listened to a tape of Fehmi describing different types of space—sitting in a chair and being surrounded by outer space, for example, and imagining the space in my head and chest. When I held the image of space in mind, the strobe and sound flickered abundantly. After a few minutes I jerked involuntarily, the way people twitch as they are starting to fall asleep, to the point that I almost fell over. That happened two or three times in the course of the half-hour session. It was, Fehmi said, stored muscle tension letting go. After the session I was dizzy for an hour or so, again the effect of the release of stress, according to Fehmi. Beyond that I did not notice many changes after a single session. After using two Open Focus tapes at my home, without the electronic equipment, for several weeks, however, the results were dramatic and among the most profound I experienced during the course of research for this book. After each session (I did two a day) I would feel completely relaxed, in a way I hadn't felt before. I could feel muscles in my head, face, neck, and scalp letting go of chronic tension, much of it presumably from sitting and typing at a computer much of the last fifteen years. Muscles around my eyes and jaw started letting go. After a couple of weeks I started noticing changes similar to those Fehmi had described. A lighter feeling. More laughter. I felt more at home in my body and at times moved more fluidly. I had more energy during my noon pickup basketball games. Some days there was an extraordinary feeling of being centered and calm, and my sense of smell was greatly enhanced. I could pick up the smell of lilacs blooming half a block away. The sun seemed more golden, richer somehow, the way I remembered it glowing from childhood. It was the first time in many years that I had felt as good. After a while the changes became part of everyday life, integrated, so they aren't as dramatic, but there has been a marked overall change in the way I feel. My account is what scientists call an anecdotal report. There are very few double-blind, controlled studies for alpha synchrony training, but the case reports are powerful. And besides, as I would find myself saying again and again about neurofeedback: what else is there?

Now in his seventies, Fehmi, who recently semiretired, has used brain wave training and coaching in Open Focus to treat thousands of clients. Talk therapy, he says, was seldom necessary, though his wife, Susan Shor Fehmi, uses the two together. As is the case with all kinds of biofeedback, some people are ready to change, or for some reason are better at it than others, and their brain simply needs a nudge that alpha provides. Fehmi described me as "ripe" and ready to respond to the training, while other people may need more hours with the tapes and brain wave sessions and gain something less. In his experience with thousands of cases, though, everyone gains something. Most make substantial gains. There is no way to independently verify that, as Fehmi has not published many studies.

Fehmi has also conducted some other kinds of brain wave training. One of the more interesting ones is yoked couples training, in which a married couple must produce the same alpha frequency in order to receive light and sound feedback. Those who do the training say it evokes a "honeymoon response," a feeling of being deeply connected on an intuitive level. Stephen Larsen, a Jungian therapist and longtime neurofeedback practitioner, did this training with his wife, Robin Larsen, at Fehmi's clinic. That night, he said, they had the same dream. "Remarkably the same," said Larsen who is also a trained dream phenomenologist. "There were fourteen points of agreement between our dreams. That's unheard of."

What I find most compelling about Fehmi's approach, though, are his profound ideas about the physiology and psychology of attention, which have been greatly informed by his decades of Zen meditation and his work as a clinician.

We gain insight into how something works when we stop seeing disparate elements as disconnected, and instead see how they work together in a system. Fehmi applies a similar systemic understanding to the human central nervous system. In many cases the widespread problems of anxiety or depression aren't fundamentally structural but rather functional, created by a stressful style of attention.

We weren't given a nervous system that would get us into stress, however, without also being given a way out. We need to step back and look at the entire nervous system, which isn't nearly as broken or dysfunctional as we think. Anxiety, fear, and depression are often not an unavoidable human condition—they are part of the misuse of the system of attention, which we understand poorly. The problems are more an operator error, a matter of how we deploy our attention, not realizing that deployment deeply affects the nervous system.

In essence Fehmi says we collect and store fear—from the time we are born—throughout our body, from our muscles to our stomach to our eyes and even to our heart. We reflexively shut down these muscles because shutting them down stops the pain at the moment. And we use what he calls a narrow objective focus as a long-term mode of attending, so as to keep from feeling pain. Narrow objective focus is a strategy to tense the body and keep it from feeling pain.

The problem is that we get stuck there, holding the pain in our body. Our fear and anxiety live on in our heart, chest, stomach, and eyes and other places and take a toll. This pain may not be felt; we might not be aware of them; but they don't go away—they affect us emotionally and physically. The reservoir of held fear is greatly underestimated and lies at the bottom of a host of maladies. How we carry this fear, or how it is expressed in each individual, is where genetics comes in. Some people may carry fear without much trouble, but many cannot. In some people the stress expresses itself in migraines; in others in anxiety and chronic pain; and in still others in heart ailments, alcoholism, or depression or all three. It creates a "feeling tone," a state where things seem dark or foreboding; it also creates feelings of dread and angst. Emotional stress is also at the bottom of attention deficit disorder, ADHD, stuttering, and obsessive-compulsive disorder. It causes or exacerbates chronic pain. It destabilizes the entire nervous system. A kid with ADHD, Fehmi's model holds, gets up and acts out to keep from feeling the discomfort of fear and anxiety. Children with ADD dissociate from the fear, or drift off and can't pay attention. According to Fehmi, the nervous system is a pot on a burner and the flame is stress. Reduce the stress—the flame—and the pot stops boiling. If the held fear can be ameliorated, ADD or anxiety or stuttering may not even surface.

And the pain can be released in a more open, diffuse style of attention. Life without fear, even if we are not aware of it, is a very different place.

Moreover, although this narrow objective focus—where we spend most of our time—keeps pain to a minimum, it also separates us from the world. It hardens us, physically and emotionally. It was designed by evolution, I believe, to allow us to function in an emergency. The opposite of "objective" is "immersed." But because the pain would rise up if we moved out of the narrow objective focus, we seldom visit the diffuse, immersed end of the spectrum. So our pain, and our unconscious resistance to feeling, separates us from a completely different and very beautiful way of experiencing the world around us.

Fehmi's exercises allow his clients to begin to slowly open their attentional aperture, to release their held fear and experience the world in new ways. It is truly an unusual approach, even in the world of neurofeedback. To my mind, it is also a more holistic method for treating problems, one which more fully recognizes that the mind is in the body as well as the brain.

Open Focus is an extraordinarily hopeful approach—it says that deep love, transcendent feelings, and the end of feeling alone and isolated and anxious are part of our repertoire, just a short journey from where we are now—we've just forgotten how to get there.

Jim Hardt, who has practiced his own unique brand of alpha and other frequency training since he worked with Joe Kamiya in the 1970s, and who runs something called the Biocybernaut Institute in northern California, says that alpha training is the antistress. Alpha engages a body wisdom, he says, an innate healing state latent in humans that fights many of society's common maladies. "Alpha's like a silver bullet against anxiety when it's applied properly," he says. Hardt has treated thousands of people with alpha training. "The physiological changes are legion," he says. "One we call the alpha tan. People look like they spent a week in Barbados. The increase of blood flow is so profound, the face is flushed. There's better sleep. Better ability to digest food. Better moods. Some people have profound changes in blood pressure. Changes in breathing. Hands and feet are warmer. Alpha training also tends to normalize muscles. " Hardt's approach is very different from Fehmi's. Back in the early days of brain wave training, he was hooked up to a biofeedback instrument, and forgotten about for several hours when the clinician went out to lunch. Hardt said his experience was profound and dramatically showed him the power of brain wave training. Now he works with clients for seven days, twelve hours a day, spending most of that time in an alpha state. He does a series of careful measurements, pre and post, that look at everything from heart rate to respiratory rate to mood and the EEG, and says the differences in measurements can be dramatic. The changes, he says, last. His intensive brand of personality makeover on a million dollars' worth of state-of-the-art equipment does not come cheap. It's $9,000 for a week-long session. Though the price tag sounds outrageous, "compare it to the cost of twenty years of psychotherapy," says Hardt.

Other highly regarded research regarding alpha training is by Dr. J. Peter Rosenfeld, a professor of psychology at Northwestern University in Chicago and a veteran of brain wave training. Rosenfeld built on brain imaging work at the University of Wisconsin, which shows that the left prefrontal cortex, directly behind the forehead, governs the positive emotional circuitry and that the right prefrontal cortex governs negative emotional circuits. Both operate in the alpha range. If the right becomes dominant and the left underpowered, it causes an asymmetry that leads to depression. The left side simply can't generate the wattage to engage positive emotions. Rosenfeld's "alpha asymmetry protocol" teaches the client to bring up the left side and turn down the right. Though it hasn't been tested yet in controlled studies, clinical reports are very positive. "It seems to work for every kind of depression except bipolar," said Rosenfeld.

Another aspect of the Peniston protocol is the theta state, which has also been the subject of much study, though it is more difficult to study than alpha because a person in theta is less conscious. The alpha state has different characteristics than theta. Alpha is a deeply restful state. There may be some emotional content or some imagery, or there may not be; there was not in my case. But as the user sinks toward the deeper theta state, imagery starts to become more prevalent.

Theta has been shown to be a hypersuggestible state, where the Peniston-Kulkosky rescripting apparently takes place. And it has other interesting properties as well. It has long been associated with bursts of profound creativity. Problems, people say, often quickly resolve themselves there in flashes of inspiration. Thomas Edison used a crude kind of biofeedback. He reportedly used to sit in a comfortable chair and hold a rock in his right hand. On the floor beneath his hand he placed a tin pie plate. He would settle into the chair, and just as he dozed off, his hand would let go of the rock, which would fall on the pie plate and wake him up. Then he would scribble down ideas that had come to him. Chemist Friedrich Kekule had reveries in a similar state in which he envisioned atoms forming a chain and snakes biting their tails, which led him to discover the shape of the benzene ring.

Elmer and Alyce Green studied theta when they discovered that every time Elmer, a veteran meditator, slipped into a creative state his EEG was displayed in theta and was accompanied by the complete relaxation of his hand. They rigged up an ingenious biofeedback device, a finger ring that had a mercury switch in it. The subject sat in a chair and relaxed. If his or her arm fell more than 20 degrees in either direction, the switch set off a door chime that brought the subject back toward consciousness. People who spent time in the Greens' lab in theta state reported images of wise old men, tunnels, staircases, and caves, all classic archetypes of a descent into the subconscious.

These deep states of alpha and theta have been known about for a long time and readily accessed and used by many other cultures. Anthropologists and other experts believe that shamanic rituals and other prehistoric beliefs used dancing, singing, and other sleep deprivation techniques to induce alpha and theta states—though they went by other names—and provided access to the unconscious for religious and personal growth, healing, and other reasons. In some rituals people would lie down while drums were played in theta frequency rhythms, while in others they used body postures or fasting or deep breathing or prayer. Meditation, of course, is the most familiar example. In the 1970s Jim Hardt hooked Zen monks to EEG instruments. He found that those who had meditated for many years had very different alpha than the average person off the street, high in amplitude and synchronous. The longer they had meditated, in fact, the more synchronous their alpha became.

One man who believes deeply in the creative power of deep states because of a personal experience is Eugene Peniston. In fact the idea for his protocol came to him, he says, when he was in theta during a workshop in Topeka with Elmer Green. "While I was at the Menninger clinic—I'm gonna share something with you—I had a hypnogogic imagery," Peniston told the crowd during the keynote address at another AAPB convention in 1998 in Orlando, Florida. "During this experience while I was in theta, I had such a vivid imagery that it scared me half to death. It was as real as we are here, looking at each other. And I refused to go back down again. And it was due to Elmer Green having a long talk with me and saying, 'This is not unusual for some people. Something is going on here, and you have to go back.' The next day I told him I would go back down. Well, it didn't take long, and I was back in the theta state the next day. And it was a continuation of the first imagery, and it didn't frighten me as bad. And then I had the third one and a fourth one, and I knew then and there what was happening to me. And the fifth one I knew it was a creative experience that I had encountered. And it was showing me a treatment that would be effective. And this is the treatment of the Peniston-Kulkosky protocol."

Whether Peniston's glimpse of the beyond was a prescient vision or a grand illusion will be decided by more research.

CHAPTER NINE

The Far Shores of Neurofeedback

∞

On any given day at the New Visions charter school in downtown Minneapolis, children can be found crawling on the floor, climbing a ladder, or teetering across a balance beam. Later they will hook up their brains to an EEG machine and play Pac-Man. It is neither recess nor physical education but arguably one of the strangest approaches to remedial reading in the country. Bob DeBoer is the founder and director of the New Visions school and a creator of what is called the Boost Up reading program. A former high school guidance counselor, DeBoer watched as many kids passed through his office on their way to graduation with no more than a sixth-grade reading level. "All I could do was be empathetic and pass them on," he says. He assumed there was nothing that could be done. But in 1980 DeBoer found himself forced into an intensive search for ways to change and heal the brain, for the same reason the Othmers and other people were forced to consider alternatives—through problems in his own family that traditional medicine could not address.

Jesse DeBoer was oxygen-deprived at birth and suffered severe brain damage. By the time she was three, DeBoer says, she drooled continually, had a vacant-eyed stare, and could only make ten wordlike sounds. "She was," the silver-haired, mustachioed DeBoer says, "a very hurt little girl." Doctors said there was nothing they could do to heal her and recommended that she be institutionalized. DeBoer, however, could not accept that diagnosis. His research led him to a therapist in Pennsylvania named Art Sandler. DeBoer took his daughter Jesse to Philadelphia for an appointment with Sandler to investigate what is called a neuropsychological stimulation program.

DeBoer learned that children have three primary pathways into the brain: visual, auditory, and tactile. As an infant is exposed to sound, for example, neuronal pathways in the brain develop that govern different aspects of hearing, a little at a time. When there is brain damage, those pathways do not form or form improperly. Because function, says DeBoer, builds this structure in the brain, Sandler's program puts a patient through a long and intensive period of therapy that uses repeated movement and sounds and visual exercises to slowly create new pathways in the brain where there were none and to take over for damaged or undeveloped areas. "If you do something often enough you're going to build connections," DeBoer says. "You're creating a pathway where there wasn't one. The analogy is that you chop down trees, then it's a pathway, then a trail, then a dirt road, then a gravel road, then a four-lane highway." DeBoer brought his daughter home, and for three and a half years sixty volunteers came into the house six days a week. At any one time three people worked with Jesse all day, three people at once putting her through motions that simulated crawling, spinning, and climbing on monkey bars, all as she lay in her bed. Sandler's work borrowed some from a controversial technique called "psychomotor patterning," developed in the 1960s by Glenn Doman and Carl Delacato.

The change in his daughter, says DeBoer, has been slow and hard-won, but she has made great gains, certainly far more than he was led to believe and more than he had hoped possible. Now nineteen, his daughter recently graduated from high school. She is not considered normal, but her IQ is 70. She rides a bike, takes her Saint Bernards to dog shows, and played three sports in school. "She functions as a learning disabled person," he says. "I was especially happy to get language. I didn't assume I would be able to, but I did."

DeBoer saw a niche for the approach in a field where medical convention holds that there is almost nothing that can be done. He and his wife, Kathy, quit their jobs and became Medicare-licensed home health care providers, treating brain damage with neuropsychological stimulation. The DeBoers saw another application as well—for children with learning difficulties. Laziness did not explain their problems, he felt; their brain and sensory pathways were stressed and hadn't formed the appropriate connections somehow. "Ninety percent of neurological development takes place in the first years of life," DeBoer says. "If it's cold or dirty in a home, or there is neglect or the baby doesn't crawl because of cockroaches, the motor pathways don't develop properly." Those motor pathways in the base of the brain play a role in the development of other senses as well. It would take more than tutoring to help kids read and learn; it would require fundamental changes in a malformed brain. The New Visions school has evolved to a program that systematically applies the stimulation technique to inner-city children who have serious reading and learning deficits. Minnesota passed a charter schools bill in 1991, and New Visions is now chartered. When this book was first published, it had 210 students, nearly all from the poorest neighborhoods of Minneapolis, and that number has now reached 300.

The neuropsychological approach is the heart of the New Visions program. According to DeBoer, some reading problems, for example, are caused by such things as an inability to hear the difference between a "p" and a "b" or to hear consonants. A hearing exam enables clinicians to pick those children out. Based on the work of Dr. Kjeld Johannsen, a Danish researcher, the children are given a tape to listen to that plays those hard-to-hear frequencies over and over again, which builds, the theory holds, neural pathways that enable the child to properly hear those sounds. The school also employs three developmental optometrists who look for vision problems—another primary factor in reading disabilities. Glasses might be needed. Or the diagnosis might call for a specific set of eye exercises. And because motor and eye movements develop together along the same neural pathways in the brainstem, the kids at this school crawl, climb, and balance on the balance beam and simultaneously look at words that are taped to floor and wall. "Crawling and creeping seems to be key to your eyes moving together," says DeBoer, "to tracking across the page."

New Visions has also had success with learning disabilities by treating something called "mixed dominance." Normal neurological development, says DeBoer, is characterized by one dominant brain hemisphere which results in a person who is either right-handed, right-eyed, and right-eared or whose left side is dominant. In 70 percent of the children at New Visions, neither hemisphere is dominant and so a child might be left-handed, right-eyed, and right-eared. "There is a competition for dominance between the hemispheres of the brain," says DeBoer. "That's an inefficiency. You can have mixed dominance and still be successful. But it makes things a lot harder. So we try to identify it and switch it over with our neuropsychological program." Though there's little research to support the notion that establishing dominance on one side or the other does anything, DeBoer says the experience at New Visions is that results of the treatment are often remarkable. "When it works, it's like a lightbulb goes on in the kid's head," he said.

Several years later DeBoer heard about neurofeedback and added it to New Visions' armory of unconventional weaponry in the battle against learning disabilities in 1991. The school now has six neurofeedback machines. "Neurofeedback and Boost Up enhance each other tremendously," says DeBoer, who believes that biofeedback makes the brain more available for learning, which allows the other therapies to be more effective. Marty Wuttke, the director of the Institute for Family Learning in Atlanta, Georgia, which has a similar treatment program, says, "Neurofeedback is the blacksmith's forge that 'heats the brain up,' and the stimulation techniques are the tools that allow us to shape it." The kids are also treated with neurofeedback for such things as anxiety, learning disabilities, ADD, and depression, which makes them much more available for learning. Many physiological problems are stress-related. Some kids—those who have been raised in an abusive environment, for example—are so fearful and have had to remain so vigilant that their field of focus narrows to a tiny aperture and they can see only a single word on the page and can't follow the words on a line. Reducing stress with biofeedback helps their field of focus slowly open.

The average student at New Visions gains a year and a half in reading abilities in a year, which is unusual in the field. DeBoer is especially proud of the fact that the program—except for the costly optometrics—is inexpensive and could be replicated in any school system. The kids at the bottom of the socioeconomic ladder, like those at New Visions, are desperate for this kind of help, he says. "There are plenty of motivated middle-class parents who will pay for neurofeedback," says DeBoer. "But the battle is getting the technology to people here. When these kids come in the beginning of the year, they are angry, turned off, or class clowns," he says. "By February they're turned on. They're trying harder. They're reading better. Everything has changed." The state of Minnesota apparently agrees. A few years ago New Visions received $70,000 to replicate the Boost Up program in two other Minnesota schools, and last year it received an additional $450,000 to take the entire program—neurofeedback and all—to twelve more schools.

The New Visions program has its critics, who claim there are few studies to show that building new pathways works. But the prevailing question raised by the story of New Visions school, and all of the places where alternatives have become integral, is What else is there?

Public schools have been graduating kids with poor reading skills for decades, for example, and they still are today. And there is nothing resembling change on the horizon. Perhaps, says DeBoer, it is time for some experimentation. Research should not have to come first, for scientists will not necessarily research a phenomenon until someone shows that it can be done. "Let us find the things that produce the results," he says, "and then we can find out how we produced it."

Neurofeedback is not being used solely as a tool for attention deficit disorder and epilepsy and addictions. Many in the field believe the technique has a far broader—in fact, an unbelievably broad—application. For example, it's being widely used by athletes, performers, executives, and others who have no problem, per se, but simply seek to enhance their "performance: to increase their memory and their cognitive abilities, to sleep sounder and wake more refreshed, to play golf or the cello better, or simply to live a better, more wide-awake life, an aspect of neurotherapy known as mental fitness training or peak performance."

There is a great deal of controversy, as we have seen, within and outside of the field, over the kinds of claims being made for neurofeedback in the areas that have little or no research to support them. Joel Lubar, one of the most careful of the neurotherapy scientists, believes that strong evidence for the efficacy of neurofeedback with ADD and epilepsy exists, and that a good deal of solid work has been done on depression and mild closed head injuries. He also believes that the Peniston protocol is a potentially powerful therapy and that early studies are very promising but that a large-scale study needs to be done. Although he believes that other applications are certainly possible, if not probable, Lubar is worried that the fringe reputation the field already has will get worse if clinicians move too far too fast and the claims once again get way out beyond the research. "They report an occasional case when they get results that look promising," said Lubar. "But they need to have follow-up data. If they don't, the person could have a relapse and we wouldn't know it. It could be a placebo. And they need a theoretical model. Why should it work with PMS, for example? I have no idea, nor does anyone else. People got phenomenal results, call it a miracle, and put it on the Internet. And that puts people off and hampers the efforts of this field to be accepted." In other words, for those in the biofeedback community, it's déjà vu all over again.

Barry Sterman, the discoverer of high-range EEG biofeedback, thinks the expansion of neurofeedback to areas beyond the research is fine, as long as the research follows. The technique will eventually prove to be a powerful intervention for all types of problems, he says. Sterman no longer works with EEG Spectrum, but treats some patients clinically. Brain wave training, he says, continues to amaze. "I treated a fellow who really wanted to be a marine," said Sterman. "But he couldn't pass the test. You need a fifty to pass but he had taken the test five times and had averaged a twelve. He had intellectual problems, memory problems, and attentional problems. He came across as if he was mildly retarded. Yet he had learned to fly a helicopter. Eighteen months after I started to train him he got a sixty-five on the test. Everything changed. Memory, attention and intellectual abilities.

"I'm excited about it," Sterman said. "I think research will show that neurofeedback will turn out to have wide-ranging application."

The Othmers, who have pushed the clinical envelope the farthest, are unrepentant. "Without the clinicians none of this would be getting done," Siegfried says. "It's the arrogance of researchers to say, 'Wait for the research.'" There is simply too much need for such powerful intervention to wait for funding agencies and universities to come around. "It works," he says. "That's all we need to know."

The Othmers have been hooking up people with such success for so long that Siegfried gets testy at people who question the claims. I was at a meeting at a hospital in Helena, Montana, when a retired neurologist stood up and pointedly questioned the Othmers and implied the whole thing was smoke and mirrors. A tense game of verbal tennis ensued. Afterward, Siegfried was seething. "I hope I'm around when that tree falls," he said. But how can someone not be skeptical when something flies so flagrantly in the face of established medical science? Here is a device that two nonmedical people claim will substantially enhance the lives of people with dozens of intractable medical problems. Yet there is also a substantial deficit of humility by some in the medical profession who refuse even to consider the possibility that the technique may work. It's an interesting phenomenon that has little to do with science. In fact, the real barrier to the acceptance of neurofeedback may be psychological, not scientific. Professionals can accept things that they understand without studies but not things that they don't.

The number of studies that have been done for problems beyond epilepsy and ADD and alcoholism are scant to nonexistent. But the anecdotal evidence, the stories and data gathered by hundreds of competent and well-trained, intelligent professionals treating thousands of people around the world for a laundry list of problems, is compelling. If neurofeedback is not making these changes, then some other powerful unknown mechanism is, and it is nothing short of revolutionary. If one keeps in mind the caveat that there are no solid double-blind, controlled, randomized studies, or in many cases no study of any kind, and that the technique is highly experimental, the widespread application of neurotherapy is well worth reporting here; if nothing else, the work hints at the incredible possibilities. And this preponderance of evidence makes a strong and persuasive argument that funding for neurofeedback should be made a priority.

The big question is, How is neurofeedback affecting the brain? There are numerous proposed models out there. I've chosen the Othmers' (for which they have borrowed freely from other places) because it is the model with the greatest scope, is the most unifying, is being used by the greater number of people, and is one they admit is constantly being revised according to what they and their affiliates learn. It is not the only approach, nor is it necessarily the best. Their nascent model holds that EEG biofeedback can treat so many disparate problems because they are really just one problem: disregulation of the brain. Since the brain is command central and runs the show, according to their explanation, once it is strengthened, running at the appropriate speed, and stabilized, it can resist a myriad of problems better than before the training. The Othmers reason that if "self-regulation" of the brain can have dramatic, long-lived effects with epilepsy, one of the most intransigent of all problems to treat, and bring about an end to or reduction in violent seizures that have wracked a human body, it is only logical that the principle of strengthening the brain should apply to a wide range of conditions, some more profound, some less. Why not try it, as long as medicine's first rule—Do no harm—is kept in mind?

Viewed through the prism of neurotherapy, human beings do not suffer an epidemic of depression, chronic pain, immune system dysfunction, addictions, anxiety, or any one of a laundry list of other afflictions. Instead, the epidemic is in hyperactive or worn-out nervous systems—buffeted by birth perhaps, infancy and childhood certainly, by a culture that encourages overwork and has created some of the most stressful places on earth: modern cities. There are only three diagnoses under the Othmer model: One, a person is so chronically overaroused he or she cannot relax, resulting in, for example, anxiety, agitation, impulsivity, and anger. Two, a person may be chronically underaroused, resulting in some types of depression, lack of motivation, and spaciness. The third principal diagnosis is brain instability. Using an automobile analogy, the lug nuts on the wheels are loose and the front end is wobbly. The driver can sometimes drive quite well, but suddenly the car veers off in one direction or another, and there is little the driver can do. Bipolar disorder (manic depression), migraines, PMS, panic attacks, motor and vocal tics, vertigo, bruxism (teeth grinding), epilepsy, and many others are all considered stability problems.

In a sense, the model is a move back to a much simpler premolecular time in neuroscience and to a much simpler time in psychology. (The notion of an under- or overaroused nervous system was first proposed by Walter Rudolf Hess in the 1950s.) The Othmers have essentially thrown out most of the hundreds of diagnoses in the *Diagnostic and Statistical Manual of Mental Disorders,* the encyclopedia of all the psychological dysfunctions, and replaced them with just three.

Nearly all health problems, this model holds, flow from overarousal, underarousal, or instability in the central nervous system. With any of the three conditions, the stressed-out brain and the rest of the central nervous system are not robust enough to manage the body appropriately and so render people susceptible to any condition to which they may be predisposed: Joint pain flares up. They can't sleep. Or they sleep too much. They get headaches. Panic attacks. Manic depression. Pain. Depression. They are anxious. Nervous. Have panic attacks. Can't pay attention. Simply treating a person for one of the three conditions can alleviate hundreds of different and seemingly disparate problems. That is why with twenty sessions of neurofeedback in the same two sites, a patient often reports that four or five symptoms diminish or disappear. The hundreds of traditional diagnoses are simply farther downstream from the diagnosis of a stressed-out central nervous system.

This iconoclastic view of mental illness received scientific support in October 1999, when Dr. Rodolfo Llinas, an eminent neuroscientist at New York University, proposed a revolutionary new theory about seemingly unrelated disorders. Speaking at a Society for Neuroscience meeting, Dr. Llinas said that there are six layers of cells in the cortex, divided into specialized regions, that allow movement, planning, speech, and response to emotions. According to a description of his talk in *The New York Times,* the sixth layer of cells provides connection to the thalamus, which is the part of the brain that takes sensory information and passes it on to the cortex. But the thalamus is more than a relay—it is a generator with special cells that set the pace for the cortical cells. They in turn feed information back to the thalamus. This loop creates the symphony of the brain, coordinating actions, perceptions, movements, and even consciousness.

Dr. Llinas studied patients with brain disorders and found that parts of their thalamus seemed to be abnormally slow, even asleep. The critical parts of the cortex that govern many aspects of a person are misfiring, because there is no thalamic conductor to set their pace. Without proper control by the thalamus, the cells in the cortex become overly excited. Because the thalamus has many highly specialized regions a whole host of things can go wrong in a very small area. If one small area—perhaps no bigger than a pinhead—of the thalamus is suffering a "brown out," for example, then because it connects to the motor area of the brain it causes the tremor of Parkinson's disease. Another small area which is very close by connects to a different area of the cortex and it causes the cells that govern pain to become excited, and so the patient is in chronic pain. Dr. Llinas also cited some types of depression, tinnitus, and obsessive compulsive disorder as problems that may be caused by a thalamus and cortex that are out of phase.

The Othmers believe Llinas's model sheds light on the mechanism of neurofeedback. They believe that when people train with neurofeedback they are training one end of that loop between thalamus and the cortex. The changes made on the cortical end project back on the thalamus, which is deep in the brain. "Changing the cortical rhythm changes the rhythm of the thalamus," said Sue. "If the cortex slows down the whole system slows down. The lower parts of the brain get involved because the system is interconnected."

The most critical variable in the Othmers' model of higher-range training is frequency. The general approach is the 12-to-15-hertz, or SMR, training on the right side, which calms and enhances the emotional aspects and relaxes physical tension, while 15-to-18-hertz training on the left side helps improve attention and alertness. In 70 percent of the patients the Othmers and their affiliates see, the approach is the same: the left side needs to be brought up to the 15-to-18-hertz frequency, while the right side needs to be calmed to 12 to 15 hertz. The reason, says Sue Othmer, is that the left hemisphere organizes more localized function, which requires higher frequencies, whereas the right hemisphere organizes function more globally, which happens with lower frequencies. While this is the general approach, each brain is different, and frequencies are tailored to each individual's response. A change in the Othmers' approach is where the electrode is placed on the brain. For years neurotherapists used only C-3 and C-4 sites, which are midway between the top of the head and the ear on each side, and some still insist that those tried-and-true sites are best. But the Othmers and others have experimented with sites on the temporal lobe, an inch above the ear on each side, for problems of stability, and they have proven to be very effective. And they have combined their training on the central parts of the brain with prefrontal and parietal sites to enhance specific effects related to the function of those parts of the brain. In other words, training right brain might bring up emotional aspects, while adding FP-1 helps coordinate the emotional well-being with thinking and planning, since FP-1 is over the prefrontal cortex, which governs those issues.

Another variable is the hemisphere of the brain that these frequencies are trained on. Management responsibilities are divided between the two spheres of the human brain. The left brain controls language, talking, reading, writing, and sequential information processing, such as keeping score, doing arithmetic, typing, and grammar. The right brain, on the other hand, perceives facial signals; governs singing and swearing; comprehends music, emotion, body language and body awareness, and environmental sounds (such as birds, bees, and babbling brooks); controls visual spatial tasks, such as throwing a football or riding a bike; and insight and intuitive reasoning. If the left side is trained, a person will see his or her analytical skills and sequential learning abilities improve. If the right side is trained, the Othmers believe, it brings up emotions and feelings, and even promotes intuition. The Othmers believe that training the left or right cortex, the top layer, strengthens the subcortical structures as well. The training doesn't so much influence the local area around the electrode, they say; rather, it influences networks of brain communication.

The Othmers' model has migrated from the motor strip to the temporal lobes to deal more directly, Siegfried says, with emotions, though it still retains the basics of arousal and stability. Much of the rest of the field use the older model.

There are a number of reasons the traditional medical model has so many different diagnoses, according to the Othmers. One of them is that there are usual genetic and environmental differences in people, and disregulation can affect people in different ways. Another is that disregulation happens along a continuum or spectrum. The mildly disregulated brain manifests in ADD or anxiety; the more profoundly disregulated brain may be autistic or suffer fetal alcohol syndrome. "The disregulation model can restore simplicity to what has become a very cumbersome process of diagnosis," says Sue.

The arousal model, the Othmers say, explains why our society is marinated in brain-altering drugs, licit and illicit. Americans spend about $55 billion a year on legal drugs. Illicit drugs, as well as coffee, nicotine, and alcohol, add billions more. Viewed from the point of view of many of those who practice neurofeedback, most of those drugs are going largely toward one purpose: to moderate the level of arousal. People who are lower on the arousal curve drink coffee to get themselves up, or if they are children, they may poke the kid in the seat next to them to get some stimulating response; while those who are chronically overaroused move themselves lower with alcohol, for example, or with something more powerful, a prescription drug such as Valium. A craving for stimulation or relaxation may be the most powerful urge of all. When the brain makes demands, few can resist.

The notion that all mental illnesses are the same but occur along a continuum of severity is actually an old idea. In 1963, Karl Menninger, a pioneer in the treatment of mental illness, wrote, "We tend today to think of all mental illness as essentially the same in quality, although differing quantitatively and in external appearance." What changed that notion was the advent of lithium, a naturally occurring element that worked remarkably well for bipolar or manic depression but nothing else. That view has been rendered an anomaly by a new class of drugs. "Prozac went in the opposite direction," says Sue. "It was embarrassingly broad in its application." It proves, she maintains, that the hairsplitting of the diagnostic model is inadequate. She sees support for her disregulation model and neurofeedback's sweeping approach in the way psychotropics and other drugs are applied. Prozac, for example, is used to treat not only depression, but also premenstrual syndrome, panic anxiety, eating disorders, obsessive compulsive disorder, ADD, and other problems. Anticonvulsants such as Tegretol, Dilantin, and Neurontin, developed for seizure disorders, are being used for everything from panic attacks to migraines, to chronic pain, to rage and borderline personality disorders. People who do neurofeedback often report that they reduce or eliminate their dependence on six or seven different medications—all being used to manage arousal or stability.

Many biofeedback practitioners have long advocated—but been widely ignored—that simply prescribing drugs or performing surgeries to treat symptoms is perhaps the greatest shortcoming of modern medicine. If 75 to 90 percent of the visits to primary care physicians are for stress-related problems, allopathic medicine falls terribly short by not treating stress. Giving someone a pill for depression or anxiety, it has been said, and not dealing with the underlying problem, is akin to taking a car that has a red trouble light glowing on the dashboard to a mechanic who simply unhooks the light.

The Othmers' model is a complete and, for a solid-state physicist and a neurobiologist who never completed her Ph.D., seemingly arrogant rewriting of the disease model. And they haven't just rewritten the model, they have said myriad previously untreatable problems are fairly easy to treat. They are in essence telling the medical world, "Put on hold some of what you think you know—from the years you spent in medical school, from years of treating patients—about what couldn't be done. We've got something here that you have never heard of that makes much of what you're doing look like horse-and-buggy stuff."

Sue shrugs as if to say, "Can't help it, that's the way it is." "The body has an incredible, innate sense of healing, and neurotherapy exploits that," she says. "Once the brain is shown the way and learns where it functions best, it will stay there. It's all about the nervous system finding its own balance. We're not changing it; we're allowing it to change itself. Acupuncture does the same thing. If you give the brain a chance, it will do what it needs to do and will heal itself. Neurofeedback allows the brain to pause for an instant, and allows it to reset. And doing that is an incredible thing to witness." Adam Crane's analogy is that stress causes neuronal pathways to jam up. The brain in that condition is like a block of steel. Brain wave training, he says, converts the steel to iron filings that reorganize along the natural lines of force.

Among the people who treat clinically with neurofeedback, the conversations are sometimes startling, almost beyond belief. I attended an EEG Spectrum training seminar on neurofeedback in Seattle. As I listened to Sue Othmer's portion of the talk, I couldn't believe my ears. Every half hour as the lecture subject changed from one health problem to the other, she would say, "Compulsive eating? That's one of the things we do best," for example, or "Anger? That's fairly easy to do," then she would tell which sites to train. During a break, we had lunch in the hotel restaurant, where our waitress greeted us with an extremely high-pitched Betty Boop voice. "SMR would bring that down," Sue said offhandedly. "SMR's effect on relaxing muscles can bring down the tone." It sounds almost preposterous, even to people who understand neurofeedback.

Indeed, it is tempting to agree with the criticism of the expansive claims. In spite of the research, in spite of the testimonials, I probably would have dismissed all of this and gone on to other things had I not experienced it firsthand in a dramatic way. When I was doing research for a *Psychology Today* story, I did about fifteen sessions of C-3 beta,

C-4 SMR. (It's fairly easy to do. Spectrum shipped me a loaner instrument, and with help over the phone, I set the equipment up and was able to start doing sessions in just a few hours.) As I started to near my eleventh half-hour session, I gradually noticed my energy level increasing. My mood was more buoyant. I started sleeping more soundly. I started waking up in the morning right away, without the usual attendant fogginess and the urge for another cup of coffee. I could work much longer without wearing out. The differences were dramatic. Yet I also felt this coldness coming over me, a kind of emotional detachment that I found unpleasant. I mentioned it to Sue. "Oh, too much left-side training," she said, immediately. Sure enough, more right-side training "warmed me up," as they say, as my emotional feelings started to return. I started turning up the frequency on the right to avoid sleepiness—from the window of 12 to 15 hertz, up half a hertz at a time. As I got up in the 14-to-17 range, it amped me up something fierce, especially right after I finished. I remember leaving the house after a session and driving to the grocery store, and my internal monologue was chattering away a mile a minute. Sue laughed when I called. "Too high," she said. "Walk the frequency back down." I did, and it fixed the problem.

So neurofeedback is off and running, being used for all manner of things. Perhaps the place where neurofeedback will make the most dramatic difference, and where it already shows tantalizing promise, is in the area of treating criminal behavior. In a previous chapter, a young man who suffered ADHD saw his impulsive, out-of-control life dramatically altered by learning to change his frequency in several dozen EEG neurofeedback sessions. Neurofeedback has been used in several other ways at prison facilities and with at-risk groups of kids, and the few studies that have been conducted show great promise.

A growing body of evidence, made possible by a new generation of brain imaging, shows that in many cases neurological damage is at least partially responsible for criminal behavior, perhaps far more than we realize. Dr. Dorothy Otnow Lewis is a psychiatrist at Bellevue Hospital in New York and a professor at New York University School of Medicine who has pursued the brain-crime link for years. She and Jonathan Pincus, a neurologist at Georgetown University, have published a series of groundbreaking studies that demonstrate a link between violent behavior and damage to the brain from abuse and accidents. The relationship began in 1976 when Lewis received a grant from the state of Connecticut to examine one hundred juvenile delinquents. She strongly believed that damage to the brain was responsible for delinquency and asked Pincus to help her by giving each of the kids a neurological examination. It was a defining moment for Pincus. At the time, he told *The New Yorker* magazine in 1997, he thought there was no connection between neurology and behavior. "After seeing the kids for himself," author Malcolm Gladwell wrote, "Pincus, too, became convinced, that the prevailing wisdom about juvenile delinquents—and, by extension, about adult criminals—was wrong and that Lewis was right. 'Almost *all* the violent ones were damaged,' Pincus recalls, shaking his head."

If neurofeedback can truly heal damaged parts of the brain that control key functions, such as the prefrontal cortex, it would stand to reason that some criminals should benefit from the technique. That was the belief of Douglas A. Quirk, a Ph.D. psychologist at the Ontario Correctional Institute near Toronto, who passed away a few years ago. He started working with biofeedback in 1959, using galvanic skin response, alpha brain wave training, and eventually Sterman's SMR protocol to treat several hundred epileptics. For the last twenty years of his life, he treated inmates at the correctional center. Quirk came to believe that certain types of violent behavior originated in a kind of epileptic burst in a deep part of the brain called the drive center. "If separate parts of this area are stimulated electrically, the animal subject responds as though it was experiencing rage, sexual arousal, hunger, satiety, or pleasure reinforcement," Quirk wrote. "If electrical stimulation of the drive center were to be achieved in ambulatory human subjects, it seems possible that the results might include dangerous or uncontrolled behavior; that is, acts unregulated by the usual organizing effects of conscious cortical processing. Depending on the location stimulated, the actions might include 'blind rages' (assaults, malicious damage?), inappropriate sexual acts (sex offenses?), unexplained over or under eating, or escalating violent or addictive behaviors. . . . Nature and accident have arranged for just such electrical stimulation in some human subjects. These are people who are subject to deep-brain epileptic events." Seizures are, of course, where neurofeedback was born and where its efficacy has best been demonstrated. Quirk identified seventy-seven convicts who he felt were victims of this kind of pathology, resulting from head injury or abuse. He trained them using a varying number of sessions, from none to thirty-four or more, using both galvanic skin response and SMR. A year and a half after they got out of prison, Quirk found that those who had received zero to four sessions had a 65 percent rate of rearrests for violent offenses, while those who received thirty-four or more sessions had a 20 percent rate of recidivism.

A small study of high-risk probationers using the Peniston protocol showed promise for a different approach. Ten men—petty thieves, hot-check writers, drug abusers—were considered at risk for drug and alcohol abuse as well as further criminal activities. They received a full course of treatment using the Peniston protocol, and a year later seven of them had remained free of the law. "They're the most impressive changes I have ever seen," said Dr. Eugenia Bodenhammer Davis, an associate professor in the Department of Rehabilitation, Social Work and Addictions at the University of North Texas in Denton, who conducted the study. "I've worked with a lot of different therapies, and I've never seen anything work like this with a high-risk group. Fifty percent is good, and that's measured right after treatment."

In Missoula, Montana, meanwhile, State District Judge John Larsen has used a federal grant to buy a neurofeedback instrument for the juvenile justice court, which deals with adolescent offenders. The same teenagers, he says, show up in his court again and again. And many of them self-medicate with marijuana, cocaine, and speed. In the first six years since he became a judge, there was an 80 percent increase in adult criminal cases. "What we had wasn't working," he said, referring to locking offenders up. "It was a revolving door. The idea is to use the coercive power of the court to get them to enter treatment. We don't have the space to lock everybody up." The program was new when I visited with him but, he said, extremely promising. It combined neurofeedback with counseling, acupuncture, and a number of other interventions. Only one young man had been through neurofeedback training at the behest of the court, but it created a dramatic turnaround. "He was angry and on the run when he came in," said Larsen. "He was a very different boy when he came back."

Another approach to treating violent offenders involves a model of emotional development known to professionals as attachment disorder, a pathology that begins as soon as a child is born. For critical normal development of the brain, an infant needs what some professionals call a somatosensory bath and others call love—hugging, talking to, rocking, cooing, holding, and other kinds of personal contact with a parent or other caregiver. Such affection apparently builds robust connections between cells in the brain and establishes pathways, especially among those that govern emotions. A deficiency in the appropriate contact can cause emotional and even physical problems later in life. In one study, brain scans of neglected children show that 20 percent of the prefrontal cortex fails to develop properly. In another study of attachment, a researcher at Harvard brought in well-bonded babies. He separated infant and mother in the same room and asked the mother not to respond to vocalizations or other attempts to recruit the mother, and he filmed the results. "Within thirty or forty seconds that baby is crying or anxious, showing all the classic signs of anxiety," says Sebern Fisher, a therapist who specializes in the treatment of attachment disorder with neurofeedback in Northhampton,

Massachusetts. "This is a baby that has reason to believe that he's going to be responded to. Imagine that neglect happens as the routine. What then happens to the baby? The arousal system keeps cranking up, but there's no response. At some point the organism can't stand it anymore. After three or four months it shuts off. The light is gone. The yearning ends."

The child becomes numb physically or emotionally in the interest of his or her own survival, and the extent of this numbness borders on the unbelievable. Children with severe attachment disorder may be so deadened physically they will break a bone and not realize it, or they will eat until they throw up because they cannot tell when they are full, or they will go out into bitter cold without a coat and get frostbite and not know it. The emotional symptoms are even more troubling. They are often sadistic and violent, torturing cats and starting fires and visiting cold-blooded violence on other people. They are aggressive and antisocial and have been dubbed "children without conscience." A high proportion of them will go on to become criminals.

Fisher, who spent twelve years as the clinical director of a residential home for severely disturbed children, believes neurofeedback is a powerful way to intervene in this pathology. "With attachment disorder there is not an adequate amount of right hemisphere development," says Fisher, referring to the emotional, feeling side of the brain. "Postmortem work or MRIs show that antisocials tend to have smaller right hemispheres, while normals have slightly larger right hemispheres. The circuits aren't live. There are fewer synapses that develop, less blood flow. If the emotional apparatus doesn't come on line in that first year [with the somatosensory bath], we don't have a technique—outside this one—that I know of that can bring it on line." Fisher hooks up and trains the right brain to stimulate the reestablishment of the networks that regulate the emotions. "You can start to get an empathic response if you train on the right side, sometimes right away. If you train too much on the left side, you get a cold, detached person, and their sense of attachment is gone." The number of cases of attachment disorder treated is no more than a handful, but changes seem dramatic. The changes are physical as well as psychological, a profound illustration of the mind-body connection.

A young woman from India, who had been raised in an orphanage, came to see Fisher for such a problem. "I remember she was sitting across from me at the mental hospital," says Fisher. "Sitting is a loose term. She could not bear to be in her skin. She was on lithium and an antianxiety agent. She was physiologically a mess. Every measure for disregulation that EEG Spectrum puts out, she met. Everything. ADD, constipation, incontinence, sleep. Every single measure. Physically and emotionally. She was skating along on the top of nothingness."

Within thirty sessions profound changes were happening. "She started to warm up. Started making eye contact, having some internal awareness. She started to have normal bowel movements. She had been constipated for years. Her lithium dose dropped. She stopped drinking alcohol, very fast." After more than one hundred sessions of SMR training, Fisher says, "everything dropped away. Her bipolar illness, her ADHD, her PTSD, her alcoholism. She started to be able to talk about how she felt about things. Now we have long talks about her history. And she is warm with me." She has a near-normal life, out of the hospital and in an apartment, where she lives with a boyfriend. She has a 3.6 grade average in college. Yet there are still serious problems. "She can't be touched. Can't stand it. And I think what we're seeing endure is the core of her attachment disorder," says Fisher. "It's still there. The proliferation of stuff around it is what we have treated."

Neurofeedback is widely used to treat depression, and those in the industry believe that given the number of different protocols—beta, SMR, alpha-theta, alpha—there are few instances in which drugs will be necessary in the future to treat depression. At Spectrum both beta/SMR training and the alpha-theta protocol are used, sometimes separately and sometimes together. While he sees his share of serious trauma, Bill Scott says people suffering from mild to moderate depression who do alpha-theta often find something simple and relatively minor at the bottom of their illness. "The kind of stuff that often comes up is a three year old who is walking through the park and the father lets go of their hand and they think their father doesn't love them. Once they've worked it out, the depression is gone." Although alpha-theta can help extract a painful memory, and there is an attendant release, the work is not always done. This is where beta and SMR training are helpful. They strengthen the brain for the work of recovery.

Those who practice EEG biofeedback use the phrase "the kindling effect." It means that even though a memory—the original emotionally stressful insult—might be remembered and processed in therapy or an alpha-theta session, it was present in the brain long enough that the wound evolved from strictly psychological to physiological and has caused problems similar to a closed head injury: short attention span, fogginess, and poor memory. Dr. Daniel Kuhn, a psychiatrist in Manhattan, specializes in treating veterans of the Israeli war of 1973. After talk therapy he uses beta training to treat the residual cognitive problems. "You can't talk people out of these kinds of things," he says. "Nothing works as well to clear them up as EEG neurofeedback."

Biofeedback is even more controversial when the subject of autism, fetal alcohol syndrome (FAS), and other similar types of brain damage comes up. At most several dozen cases of autism have been treated with neurofeedback, and many of them have turned around dramatically. Steve Rothman, the psychologist in Bellevue, Washington, who leases neurofeedback machines to people for home use, has treated six cases of autism and sixteen cases of FAS. Eighty percent of his FAS patients have made substantial gains, he says, and five of the six autism cases he treated also showed dramatic, and apparently long-lasting, gains. One of those was Rene Hockenberry's son Brock Hunter.

Brock was in sixth grade when he started neurotherapy, but his academic skills were at a third- or fourth-grade level. He was extremely withdrawn, and a change in his routine or unfamiliar surroundings would cause a great deal of anxiety. When he got frustrated or anxious, which was often, he displayed symptoms typical of autism, or, as it is called now, pervasive developmental disorder (PDD)—tics and shaking and hand wringing and strange deep sounds.

Hockenberry heard about neurofeedback from a friend and traveled from her home in Spokane across the state to Rothman's office near Seattle. Changes started to show up just two sessions into the treatment, she says, as Brock's frustration diminished. After twenty sessions, Brock told his mother that he was bothered by something he had done. "He said, 'Mom, this has always troubled me.' And I said, 'Why didn't you tell me before?' And he said, 'Because I couldn't explain it.' He started telling me that he was angry or frustrated instead of doing the tics, the hand wringing, and the guttural sounds. I finally understood what was happening. I had always been able to see the information going into his brain, and he could process it, but nothing came out. It was like the cable had come off the battery, and there wasn't enough power. Neurofeedback charged up his brain." After thirty-six sessions, the changes were dramatic. There was much less frustration and anxiety, and he started to become aware of how he felt. Now, several hundred sessions later, Brock is no longer shy or subservient around strangers, nor does he hug the wall when he walks in public. Hockenberry says he often walks up to someone he doesn't know and introduces himself. People can touch him, and he doesn't react negatively. His severe allergies are close to gone. His schoolwork has gained—he's very good in math and art, but his handwriting remains at a fourth-grade level. "He was black and white in his thinking," Rene says. "And he had no sense of humor. Now he does. He even tries to make jokes. They're not funny. But he tries."

For a while she worried that the incredible gains would disappear. After thirty-five sessions, the tics had stopped, but if there was no training for a few weeks, they would start to return. Now, two years after the training began, there is no more backsliding. "I took him to a demolition derby," says Hockenberry. "A year ago, forget it, with that much stimulation he would have been spazzing," or reacting nervously with tics and trembling. "But now he sat there calmly. He was fine."

After paying $85 a session, for two sessions a day, she says, "It cost seven thousand dollars and cleaned out my savings and my inheritance." Though it was well worth the cost, she needed to find an alternative. She and her husband, Brett, took out a second mortgage on their home and bought an instrument of their own; Hockenberry now treats her son.

Hockenberry says Brock has gained far more than she ever imagined. She will keep on training him and firmly believes that she "will make him whole." He still speaks in a monotone, and she would like to change that. And she would like him to focus his extraordinary memory on more than television. Will his jokes ever be funny? She laughs. "I certainly hope so."

Testimonials don't necessarily add up to science, of course. Toronto practitioners Michael Thompson and his wife, Lynda, have treated several clients with PDD and Asperger's syndrome, which is a problem similar to autism though not as severe. They both caution that the approach at this point is highly experimental. "We know we can help the symptom picture and the attention span. That's part of the picture," Michael says. "But autism is a much bigger problem. We can only have conjecture about the results we've had with autism. The numbers are too small. But that's how things start. We shouldn't be ashamed of anecdotes, but we should be very cautious." "You won't get normalization," adds Lynda. "You don't want to raise hopes that this could be a cure-all. But if I had a child with those symptoms, I would certainly try it."

Rothman expects skepticism about such claims. But he says that results he obtained with Brock and the others he treated are what he fully expected to happen. And that ability to predict what will happen is key. "The things [Rene] was reporting were extremely consistent with what we would expect to happen with this technology," he says. "Better sleep. More vocalization. Less shyness. You can predict outcomes with this model, and for me that's a compelling argument for its efficacy."

After years of research on neurofeedback, including my own experience and the experience of family members, I am no longer skeptical about whether it works. Instead, the questions are about the scope of the success: What percentage of people respond dramatically and successfully? Miracle stories are fascinating, but how often do they happen? How many people get something out of it but not a transformation? How many don't respond at all? How long does it last? Can neurofeedback cause problems? If the efficacy question has been, with few exceptions, poorly researched, these kinds of questions have barely been touched. All that exists are anecdotal data.

Margaret Ayers, having treated thousands of people with her own hands, has more clinical experience than anyone else, and estimates that more than 90 percent of the people she treats see complete or nearly complete relief from their symptoms. Joel Lubar claims that more than 90 percent of his clients gain substantially from the technique. The Thompsons see a success rate of 80–90 percent. Rothman estimates that 70 percent of the people he sees get substantial help with neurofeedback; that number is closer to 90 percent, he says, with people who have fetal alcohol syndrome or autism. Sue Othmer, whose clinic has treated more than two thousand people and who is party to data on those who are treated by hundreds of Spectrum affiliates, estimates that 20 percent of the people she has treated are miracle stories. "These lives are on a completely different track," she says, "and people know it." As an example she cited a near-drowning victim who went from being rigid, hysterical, and in extreme pain to being a calm, relaxed child with increasing awareness of his life. An additional 45 percent, she says, gain substantially and get what they came for: normalized sleep and attention, better mood, less depression, and reduced or eliminated pain. Twenty-five percent get some improvement out of training but not all that they had hoped. The response of that group of clients, Sue believes, is complicated by allergies, an unstable home situation, or nutritional or endocrine problems. Less than 10 percent of the people Spectrum treats, she says, are not moved at all, perhaps because their brains are somehow hardwired differently, or because they fear or resist change. All clinicians say that in all but a few cases the training lasts for years, though occasional booster sessions might be needed. Some say that chronic fatigue syndrome and epilepsy have a higher rate of backsliding than other problems and might require more sessions.

And what about problems caused by neurofeedback? Unlike miracle stories, they are far more difficult to extract from practitioners. People do not offer them up as readily as they do the tales of transformation. But there are some. "In my mind, the danger is simply that it is a very powerful technique, and most people don't know that yet," says Sue Othmer. "So they say it certainly can't hurt you. And strictly speaking we can't hurt you, but we can certainly put you in an uncomfortable place. Just like coffee or alcohol can put you in an uncomfortable place." But a key concept of neurofeedback, according to the Othmers, is that anything that can be done can be undone. "We have an antidote for everything we do," says Siegfried. Neurofeedback's effects happen gradually enough that the patient and therapist can usually realize what is happening and change the treatment.

The Othmers have had a couple of specific problems they told me about. Because self-reporting is key to knowing what's going on with a patient, those who are not tuned in well to their bodies or are uncommunicative pose problems for a clinician. One man came to Spectrum with trauma to the right brain. His wife had insisted he go, and he was not willing to communicate how he was feeling. They trained him on the right side at a high frequency, and, Sue says, "We got him really speeded up. And we didn't know it. He didn't tell us. He drove home at rather high speed and terrified his wife and started acting really bizarre at home—taking everything out of the garage and putting it on the lawn. She called me and finally convinced him to come back. He drove back, again at high speed, and we did SMR on the right side and calmed him down."

Another man came in for back pain to see if he could head off surgery with neurotherapy. "This fellow we actually moved into having migraines," says Siegfried, "something he claimed he hadn't had before. And it took us some while to get him back out of the migraines, even though migraines are something we generally remediate quite easily." Just as people can have their seizures reduced, they can also be made to have more seizures if the protocol is incorrectly applied.

Alan Bachers, a psychologist in Chagrin Falls, Ohio, who has been using biofeedback for twenty-five years, first as director of the biofeedback program at the Cleveland Clinic and then in private practice, where he treated a family friend for a stroke. This woman was depressed, had aphasia (a loss of speech), and was agitated and nervous. She had been a high-functioning individual who worked as a real estate executive. After a couple of dozen sessions, she started to change. "We made great gains in speech," Bachers says. "She could make complete sentences. She could socialize. She was much less depressed." Then, boom, one day she was worse than she had ever been. Did she have another stroke? Did Bachers do something to cause the problem? He doesn't know, and that in itself points to a problem. With help from Prozac and more training, she has made some gains, but it could be a long time before she gains back what she had earlier.

Early on in her career as a neurotherapist, Sebern Fisher was treating a young woman with an eating disorder. "She started having raging eruptions," Fisher says. "And she started coming in and telling me she was stealing, a new behavior. She started saying things like 'I don't care. I don't care about these people.' She became cold-hearted and disinterested. A whole persona developed." Too much left side? "Quite. I just went entirely to the right. She warmed up again. And she doesn't steal, has no desire to steal. I don't think you make people into something, but you bring out something in them."

Another, more subtle problem with brain wave training, says Fisher, is state change. People change, but the change is gradual enough so that although they change, they don't realize it. This is especially a problem when people are self-training. As neurofeedback gradually starts to move the brain into a new place, people can feel very different. Often they feel better, extraordinarily so, but sometimes its not who they really are. Fisher offers a benign example. She came home after training herself and said to her husband, "We have got to clean this house now." He knew it was the left side training, but cleaning the house immediately seemed perfectly logical in the pumped-up, aroused state. "The state," she says, "creates the narrative." This is a key criticism, in my opinion: people doing high-range neurofeedback can get into a state that may lead them to do things they wouldn't normally do—have an affair, for example, or buy unnecessary things. The problem is that in this state it seems perfectly logical.

Another neurotherapist, who asked not be identified, treated a fourteen-year-old boy who had been in an accident that had shattered his skull, which required a plate and left him in a coma for a month. When he came out of the coma, he was weak on the left side, so the therapist began training on the right side of the brain, which governs the left side of the body. "After a session he claimed I had sexually molested him, and all this bizarre craziness came out of him," the therapist said. "The right hemisphere is where the emotional stuff is lodged. I don't know whether it was sitting there, and we stirred up the memory or what. It scared the shit out of me," said the therapist, who is waiting to see if the county attorney will file charges. "Neurofeedback is uncharted waters. Every brain is different. Every injury is different. When you work with the brain you never know what you are going to get." By and large, the problems so far seem to be minimal and correctable. There have been no malpractice lawsuits. But there is still a relatively small sample size, and serious problems may yet appear.

A different kind of problem for therapists is that some people change so gradually, they sometimes don't realize they have gotten better, even if the change is dramatic. "They'll say, 'I don't really feel different,'" says Sue Othmer. "I'll say, 'How about your headaches?' and they'll say, 'Yeah, they're gone,' and I'll ask them if they're sleeping better, and they'll say, 'Yeah, I guess I am!'" They feel more like themselves and forget what it was like to feel bad. Most therapists do a symptom checklist so their clients can compare in detail their condition before and after training and not forget how bad they felt.

Not everyone who gets into neurofeedback stays with it, for various reasons. One problem is that neurofeedback selects for the most difficult cases. People whose problems are easily remedied with drugs or cognitive therapy usually don't seek out alternatives. Those who have more serious problems end up seeking alternatives, and so brain wave trainers end up with cases that are harder to treat. "By the time people are willing to try alternative therapies they have already tried the gamut of mainstream therapies," said Deborah Pines, who is a talk therapist but no longer a neurotherapist, but does solely talk therapy in New York City. "They put all of their hopes and expectations on this cool, newfangled treatment. When it doesn't work, they are more disappointed. They would leave frustrated. People realize therapy doesn't always work, but for some reason with neurofeedback they have higher expectations."

Another concern about higher-range training is a philosophical one, similar to concerns about a nation that finds happiness in a pill or other drugs. If people can hook up to a biofeedback instrument and make themselves feel better without ever having to delve into their past, have they bypassed a chance to learn something about themselves, forgone an opportunity for personal growth? Are parents who take their kids into a clinic for forty sessions to treat ADD or depression making up for neglect or another kind of bad parenting? Will society ever be forced to make the kinds of changes necessary to head off crime and child abuse and violence if we have a way to fix things ex post facto? Does it matter? Or is simply feeling better most important? What are some of the other implications? Will neurofeedback lead to social engineering, a *Clockwork Orange* future, where some people are forced to do it to make them more manageable and less violent? Is that wrong?

One of the major concerns among some in the field of biofeedback is whether there is adequate regulation of biofeedback, and how "adequate" should be defined. Are people trained well enough in the Othmers' five-day workshop, or anyone's workshop, to hang out a shingle as a brain wave trainer? The Othmers fear that neurofeedback is such a powerful intervention that once the establishment gets over its initial disbelief and realizes what it can do, the technique may be medicalized—that is, taken out of the hands of just anyone and restricted to use by medical professionals. There is virtually no government regulation; right now the Othmers regulate themselves, and they have been limiting who may buy the technology. It used to be that anyone could buy a Spectrum instrument. Not anymore. A neurotherapist must now be a professional licensed in their profession—a counselor or psychologist, for example—who has taken the six-day professional training course, and who has malpractice insurance. People can buy instruments for home use, but even for mental fitness training they must work under the auspices of a professional. Joel Lubar also restricts his training to licensed mental health and medical professionals. But some other manufacturers have no requirements at all about whom they will sell equipment to.

Chris Carroll argues that licensure is critical for fee-for-service biofeedback providers, and that licensing should be created in each state for biofeedback. Medical professionals, he also believes, need to be involved in the therapy. "I don't think anyone should be doing biofeedback on any patient who is reporting physical symptoms without a full medical review by a physician," Carroll says. The problem is not the effect of the biofeedback. What if, for example, someone has headaches and comes for treatment, and those headaches are caused by a tumor? "If you have people who don't recognize the risks of keeping a person from visiting a doctor, that scares the hell out of me." Biofeedback, he says, should be practiced in the same way physical therapy is: as an adjunct to medicine. Doctors and psychiatrists should refer patients to a biofeedback therapist.

Carroll is involved in efforts in New York to pass a law to establish a licensing process for those who practice biofeedback. Biofeedback has always been viewed as a relatively mild intervention, and right now anyone can buy equipment and legally practice brain wave biofeedback with no training of any kind. The Biofeedback Certification Institute of America, in Wheat Ridge, Colorado, has a training course and exam that leads to certification for both traditional and EEG biofeedback. But it is strictly voluntary. There is no government control over who can practice biofeedback at this point. "Licensing would define the scope of practice with clear ethical guidelines," Carroll says. "People would have to meet specific criteria to qualify as biofeedback practitioners." Licensing would also establish a board that would enforce the standards.

"Establishing practice standards is vital," says Alan Strohmayer. "In the absence of a systematic research program, all you have is anecdotal data. But if we can't rely on the people who are reporting the anecdotal data, then we're really screwed."

While neurofeedback serves as a treatment for any number of problems, it is also widely used by people who have no problem per se but simply want to enhance their cognitive skills and creativity, have more energy and perform better, sleep better and improve memory. A cross section of sports and other high-performance professions are beginning to use the technique.

Rae Tattenbaum has an advanced peak performance program that utilizes brain wave training in her office in West Hartford, Connecticut. A clinical social worker, Tattenbaum worked for ten years as a specialist in the use of technology in education and health care for Northern Telecom. She got so interested in neurofeedback that she quit her job a few years ago to devote full time to it, combining it with her other interest: singing.

Tattenbaum will treat anyone to optimize what he or she does—she has worked with photographers, kids involved in sports, actors, a speed skater, and gymnasts—but she specializes in training opera singers to enhance their performance, and she teaches children from the Voice and Theater Department at the Hartt School in West Hartford, Connecticut. Tattenbaum borrows from several practitioners to create a treatment tapestry that clients say works very well. She starts out with twenty sessions of SMR 12-to-15-hertz training on the right side, which centers and relaxes a person and, she says, strengthens the ability to be in the present, which keeps performance anxiety down. Training 12 to 15 hertz keeps them more firmly planted in that frequency and less likely to drift accidentally into theta. "The theta range [the subconscious] is where the fear of performance lies," Tattenbaum says. "That's where childhood fears are. I don't let them go there. I keep the inhibit tight, to keep them from drifting into theta. And because of that, they can focus better on what they are doing." The SMR, again on the right brain, which governs creativity and the imagination, allows people to more vividly do guided imagery. Tattenbaum has designed elaborate guided imagery exercises, based on the work of the psychotherapist and author Nancy Napier, so performers feel safe when they perform. Because working on the right brain causes a rise in emotional feelings, she says, the SMR brings up feelings of empathy. "We're opening them up to be able to get in touch with the passion of the song," she says. The training also improves the singer's ability to discriminate among sounds. She often does alpha-theta training as well, which releases traumatic childhood memories.

Tattenbaum also uses Les Fehmi's Open Focus tapes, adapted especially for singing and designed to relax the groin, the diaphragm, and especially the throat and the facial mask. "All the peak performers I've worked with have this facial tension, and they have to work at letting it go. Once they learn to relax the face and their focus, it relaxes the whole physiology," Tattenbaum says, "and they feel like they own their space." Coupled with the calming effect of the SMR, the Open Focus "helps them to bring the throat to a relaxed state and better stimulate the diaphragm." They seem to have more range. And if all that weren't enough, she says, the kids she works with also do better in school. The cost of a course of treatment is about the same as a set of braces—from $2,500 to $4,000.

Mitchell Piper, the former chairman of Voice and Theater at the Hartt School and now an agent, has worked with Tattenbaum for the past thirteen years and believes the technique has done wonders for the two dozen or more students he knows who have done the training. "I've seen great success with the students I've sent to Rae," he says. "I've seen their vocal ranges expand. Their ability to sing a vocal line improves. Their ability to communicate a text improves. The colors in their voice open up. All of these things are what we're looking for. I've not seen any students who have not benefited."

One singer who benefited from the training is Robin Blauers, who also taught at Hartt. Forty-two years old, she has been singing since she was eight years old. "I noticed that I couldn't be present as a performer," she says. "I used to feel that I was outside myself, watching and judging. I could never fully be present. It's like driving a car. You arrive at your destination but on automatic pilot, and you don't remember driving there." A year of training, she says, changed that. "I did a recital last July, and it was the first time I really enjoyed myself because I was present in the moment. The training gave me an overall sense of well-being that I didn't have before, whether I am singing or at work. I can step back, take a deep breath, and focus on things much better."

Many of the things that neurofeedback claims to do are the things that athletes have long sought to enhance their performance. "The most important aspect of athletic performance is control of the physiology during competition," said Dr. Vietta "Sue" Wilson, a professor of kinesiology and health science for the last twenty-eight years at York University in Toronto where she trained archers, basketball players, track and field, wrestlers and many other athletes, most at a national or Olympic level.

She has been using traditional biofeedback for much of this time. For example, she will outfit a skater with a portable heart biofeedback monitor. Heart monitors had been used to train racehorses, so a jockey knew when the horse was at the target rate for the most benefit. In sports a soft beep lets the athlete know they are in the range where they perform best, and teaches them what it feels like to be there.

Wilson has used quantitative EEGs to study performance and has used neurofeedback to train athletes. But she sees an application for high performance in several ways. Anxiety is one of the biggest enemies of an athlete. A ski racer, for example, falls behind and starts a cascade of negative thoughts. He worries uncontrollably that he won't be able to win, which in turn causes the heart rate to accelerate and adrenaline to surge, which causes more anxiety and takes the athlete out of his game. "A change in brain chemistry during performance changes the speed of mental processing," Wilson says. "It changes your decision making. You either go too fast and make errors or you go 'what if, what if, what if.'" Other people start looking at consequences and they start to tighten up in the muscles. They can't focus. You will see rushing and poor decision making." A basketball player will overshoot or take a bad shot. Working with top rifle shooters, she has noticed one element she believes is the essence of peak performance. "A stilling of the left brain, which governs self-talk, right before the shot," she says. "It's key. You want to calm the mind and see less chatter." Neurofeedback does that well, she says. "I don't believe in being relaxed when you are performing. You are aroused, and you should be aroused, but only in the appropriate manner."

Professional golfer Ed Galvan says that neurofeedback dramatically changed his golf game. The forty-year-old Galvan, of Alhambra, California, is a self-confessed golf addict who started competitive play in 1987 and plays regularly in tournaments in Asia, Australia, and southern California. He raves about the training, which he did with Bill Scott at EEG Spectrum. It has taken three strokes off his game, he says, and was the reason that he was ranked in the top twenty in the southern California section of the PGA of America. It hasn't changed his income from golf, he says, which was $26,000 in 1998. But that doesn't matter. "It's eliminated the head trips and taken my game to the next level," he says. "I'm definitely much more relaxed. When you play golf, too many thoughts come into your head—doubts about your skills or where the troubles lies, a lake or out of bounds. What EEG [training] does is develop the brain to maintain a thought for a longer period of time. I can focus on the target line, where I want to hit the ball. It's turning everything else off and being in that one line. It's a whole new level of concentration. People say they are concentrating, but I don't think most people know what concentration really is."

Bill Scott says the neurofeedback protocol for each athlete differs, according to their problem. "If it's more about focus and concentration," he said, "it is twenty sessions of beta and SMR. If it's about technique or it's a psychological issue such as confidence, then it's a majority of alpha-theta." Sometimes it's a blend of both.

A question many people ask about peak performance work is whether neurofeedback can encourage the feeling of what Abraham Maslow called "flow," or peak experience. Many athletes call it "being in the zone" and it is those times when someone such as Michael Jordan cannot miss a shot. "We can, I believe, train the conditions that encourage it," Wilson said. "Does it happen all the time? No. But it is more likely to happen."

Drs. Michael and Lynda Thompson of the ADD Centre in Toronto taught Bruno Demichelis, the head of sport science for the professional soccer team AC Milan, the basics of neurofeedback at a weeklong workshop. Michelis, with the resources of a famous soccer team, set up what he called a secret weapon: a mind room. Forward Alberto Gilardino, defender Alessandro Nesta, and midfielders Andrea Pirlo and Gennaro Gattuso all trained in the room. They lay down on reclining chairs, their bodies fastened to a device that measured seven indicators, including brain waves, muscle tension, and heart rate.

With feedback Demichelis taught them how to quickly reach a relaxed state, and how to gain control over all measurements of arousal, especially heart rate variability. Then they were taught to maintain this state of low physiological arousal while visualizing their athletic performance. If something goes wrong and a kick is missed, for example, Demichelis trains them to move out of their anxiety and return quickly to a relaxed state. He does it by showing them a video of their mistake. When their arousal levels increase in response to a mistake, they are taught to relax again. Eventually they gain physiological and mental control. Instead of being nervous and distracted in the critical minutes before a penalty kick, they can, because of their training, become relaxed and focused. Whereas neurofeedback more or less automatically calms the body, this training gives more control over a range of physical responses, including heart rate and muscle tension. "They loved it," said Lynda Thompson. "Focus indeed makes a difference in sports." They won the European championship in 2007, and later announced their "secret weapon"— neurofeedback.

In 2008 in an interview on ESPN Los Angeles Clipper center Chris Kaman revealed that had been misdiagnosed for ADHD and was overmedicated with Ritalin. Instead he had what his clinician, Dr. Tim Royer, a neurosurgeon, said was an anxious brain. Royer designed a neurofeedback program for the 7 foot tall center, and after several months Kaman made his announcement. He wanted to tell the world about neurofeedback, he said. It made him much more focused and gave him the ability to sustain his attention to hear what the coach was saying or on driving the lane.

After the training Kaman had his best season, his fifth in the NBA, and was ranked among the league leaders in rebounds, blocked shots and something called "double-doubles" or being in double figures for both rebounds and blocked shots in a given game. He was also one of the leading candidates for the league's most improved player award. "I don't want to go against doctors or psychologists or neurologists and say they don't know what they're doing," Kaman said in his interview about his experience with neurofeedback. "Because I'm not as intelligent as they are and I don't know all that stuff. But there are alternatives."

Mental fitness training is also finding a home in the executive suite. Richard Schroth is a Fortune 500 management consultant and futurist who travels the globe talking to the CEOs of large companies about information technology. His wife, Robin Moore, is a counselor. She discovered neurofeedback, attended some training sessions, and mentioned it to Richard. After a training seminar, the Schroths bought two Spectrum systems and trained the whole family for issues relating to attention and focus. The results were very positive. His family's experience, Richard says, demonstrates that the applications for stress management and peak performance in a corporate setting are phenomenal.

When I talked to him, Schroth was in the process of incorporating brain wave training into his seminars and was still figuring out how to craft it into stress management, which he sees as a growing need. Electronic commerce has dramatically sped up the way people do business, he says. Someone clicks on a Web site, an order is immediately processed, accounted for, and shipped out the door. "You have to have all of your systems integrated, along with the management component," he says, and in such a demanding environment, stress is becoming even more of a problem than it has been.

Schroth points to the military research Barry Sterman conducted. Under an Air Force contract, Sterman did research on B-2 test pilots, monitoring their EEGs as they performed the complicated task of flying a bomber. Those that performed the task best, and with the least amount of stress, were those whose brain waves were most flexible—that is, those who were able to go into the high-frequency, desynchronized state of beta and, after the task was complete, to move quickly back into synchronous alpha, a relaxed state. Those who didn't experience the alpha reward were more anxious, more stressed. The SMR and beta training, Schroth says, allows people to focus much better and attend to the task at hand; it increases their ability to move between states. In the future, employees involved in complicated, stressful tasks might wear biofeedback equipment all the time so they could "turn on" the higher-frequency brain waves when they need to and come to a complete rest between tasks to keep burnout low. "I've seen the massive stress that executives are under," Schroth says, "and I think we'll see corporations start to think about this."

Brain wave training technology is still very young, Schroth says, and extremely promising. "In the late nineteen-sixties and early nineteen-seventies, to perform many of these things, researchers were dealing with some pretty crude instruments, yet they were still getting some effect. As the technology has gotten better, the results are getting better. There's something called Moore's law. Gordon Moore was a chairman of Intel, and his law says that every eighteen months you'll see a doubling in the technology's performance and a halving of the price. I think this field will see a lot of changes."

Nursing homes also may be equipped with neurofeedback instruments one day, the experts say. Thomas Budzynski, one of the pioneers in brain wave training back in the 1960s, is now at the University of Washington in Seattle. He once headed something called the Ponce De Leon project in Tampa, Florida, doing "brain brightening" with senior citizens. "People find their memory improves, and that they have less forgetfulness," says Budzynski. "It allows a senior to lead a fuller life."

Neurofeedback, in the observation of some, not only changes symptoms and enhances performance, but in some cases can spark a new awareness of sorts and change how someone thinks about the world and his place in it. "I would call it awakening to who he was," says Alan Bachers of one such case. A client came in who drank often and wanted not to quit, but to cut back on his drinking. After six sessions of SMR/beta, he went out with his friends to a bar. "And he forgot to drink," says Bachers. "They were getting drunk, and he was not liking being there. It struck him like a bolt of lightning. 'I forgot to drink.' He liked to drink less and less and saw those friends less and less." Bachers says compulsive behaviors such as overeating, drinking, gambling, and shopping sometimes—though not always, by any means—melt away with neurofeedback training and are never missed. "These are things we fall into as substitute for who we really are," he says. "It only takes a glimpse of nonhabitual behavior to change things." Smoking is one of the toughest things neurofeedback deals with, which speaks to the power of the drug. Sometimes the habit simply falls away with beta training, sometimes not. "About fifty percent will say to themselves, 'It's not for me anymore,'" Bachers says. "But the other fifty percent are more classically addicted. The substance, nicotine, is an antidepressant, and it's tough to quit."

There are also a few of the old-time alpha guys—Jim Hardt and Adam Crane among them—who insist that the early claims that alpha training is a shortcut to the goal of transcendence and enlightenment are very real. "It's interesting that brain wave training—alpha in particular—can get people off of drugs and alcohol, can replace depression with sadness and joy, can open people's hearts with love," says Hardt. "But the single reason I have tenaciously pursued this for over thirty years is that it leads to transcendent experiences. In the early years, the rate of profound spiritual experiences was somewhere between one in ten or twenty. Occasionally, someone who was really ready would have a profound experience. While that was exciting, it was also frustrating because my mission was to bring this to everyone. As I've improved the equipment and the protocols, now we have three out of five having experiences like that." The experiences, Hardt says, include everything from direct visitations from angels and other spiritual figures, to feelings of intense bliss and joy and, most important, he says, the ability to forgive.

As neurofeedback grows, one of the most promising aspects of its future is the integration with other modalities, such as in the Boost Up program at the New Visions School. Neurofeedback is made more powerful, practitioners say, by combining it with everything from nutrition, to talk therapy, to acupuncture, to hypnosis and other therapies, some alternative, some mainstream. "It's exciting," says Paul Swingle, a therapist in Vancouver, British Columbia, and a former professor at Harvard Medical School and Ottawa University. "And we're really in the blunderbuss phase right now." Depending on the patient's diagnosis, Swingle uses everything from a blending of specific tones (played through headphones) that suppress theta brain waves, to acupuncture to craniosacral therapy, a manipulation of the plates of the skull that some believe can relieve pressure on the brain. "But the real dynamite in the package is the neurotherapy," he says.

Another adjunctive therapy is stimulation with light. The New Visions School is one place that uses a light and sound machine, a pair of headphones and goggles with an array of tiny lights inside that can play at a frequency lower or higher than that of a person's own EEG and move him or her into that state. At a low frequency, for example, the equipment can calm an agitated person. It is a somewhat controversial technique; some in the field don't like the idea of forcing the EEG and say that the technique needs more research. Proponents say it is especially useful for people who have trouble making changes with neurofeedback alone.

ADD is where many of the integrative techniques have been applied. The model for the way that neurofeedback may be used for attentional problems is at a clinic in Mississauga, adjacent to Toronto, a city of gleaming glass and stone high-rise offices. Called the ADD Centre, it's operated by Michael and Lynda Thompson—mentioned in the autism discussion earlier in this chapter—a psychiatrist and a psychologist, respectively. The Thompsons have been in the field of treating learning disabilities for a long time and are experts on the task of paying attention. Michael, who is sixty, has been treating ADD for thirty years, and Lynda, who is fifty-one, did her Ph.D. in 1979 on the subject of self esteem in hyperactive children treated with Ritalin. She is the coauthor, with pediatrician William Sears, of *The A.D.D. Book.* On a warm summer day, the Thompsons are working in their office in a pleasant, tree-lined office park, the Sherwood Towne Center, a place full of brick buildings and cobblestone walkways that resembles an English village. Inside, there are shelves with basketballs, squirt guns, and other toys. In EEG biofeedback, all users are rewarded with a visual signal—doing well on Pac-Man, for example—and an audio beeping. But that is often not enough to keep children going, especially children who have ADD and ADHD. So there is another level of reward. Children who do well in a session are handed a plastic token, and when they have amassed enough chips, they can buy any of the toys that line the shelves.

The Thompsons have five treatment rooms and use an array of different types of biofeedback equipment. They found out about neurofeedback in 1992, and it transformed their practice. "It's the single most effective tool we have," says Lynda. In their own study—not a double-blind but a simple outcome study—they tracked outcomes with more than one hundred clients, and found significant improvements in ADD symptoms and learning. Twenty-five percent of the group was taking Ritalin. After forty sessions, 80 percent of those who had been taking Ritalin were off the medication for good.

The Thompsons and William Sears, who co-authored *The ADD Book* with Lynda, advocate a careful and integrated approach to the epidemic of ADD and ADHD. Sears, in his foreword to the book, says the first critical step is the diagnosis. He sees three categories of kids being diagnosed with ADD. First are children who have what he calls situational ADD. "There's a problem in their environment, probably at their school, and it causes them to act out," Sears writes in his book. Another group doesn't really have ADD. "They are just bright, energetic, creative children who act differently and learn differently. These children are exhausting for parents and challenging for teachers, and they are just plain inconvenient for society, especially for a school system that rewards sameness and undervalues difference. These children are just differently-abled, not disabled. They have neither a deficit nor a disorder. They need a different style of parenting and a different style of learning. They do not need a label," he writes. The third group is a group that has a neurobiological problem of some sort that needs a variety of interventions, of which drugs may be a part.

That last is the Thompsons' approach in Toronto. Children are carefully tested before their treatment to figure out who they are and what kind of problem they have. Once that is determined, a strategy is devised. The Thompsons look at diet. Poor nutrition can be contributing to the problem. A Coke in the morning, or even a turkey sandwich, with its sleep-inducing tryptophan, can cause theta to increase. Poor sleep can increase theta. One boy they treated had the typical overabundance of theta. Simply by changing his sleep, his theta dropped from 65 percent of his brain waves to 23 percent, and beta rose from 16 to 33 percent. Anxiety can cause symptoms of ADD. One little girl was treated with a traditional relaxation technique, hand temperature biofeedback. Cold hands are a strong indicator of anxiety, and to be comfortably alert, a person needs to relax. Her hand temperature was 78; after some relaxation training, it was brought up to a normal 95, and her ADD was greatly diminished.

I watched as Michael treated a boy with ADD and coached his breathing, another form of relaxation training. "Take deep belly breaths," he said. Relaxation training makes neurofeedback more effective. Michael trained the boy on the F-1000, a simple computer-driven system that was hooked up to show the boy how he was breathing. It allowed the boy to try and breathe deeper, and to get a reward of a circle that grows in size and creates a sound like a UFO. Then he trained the boy with neurofeedback. "That's it," Thompson coaxed gently in his deep voice. "You want to keep your daydreaming wave down. That's this one," he said pointing to theta waves represented by a bar on the screen, "and keep your brainy brain waves up," and he pointed to the beta waves.

The Thompsons are not antidrug. (Nearly all of the neurotherapists I interviewed believe that prescription drugs have a place in an integrated approach, albeit a smaller role than is generally assumed by mainstream medicine, and in some cases as only a bridge until people can do enough neurofeedback sessions to make a difference.) Stimulants have a role to play, they say, albeit a much smaller one than is widely believed. "Meds are fine for short-term management of certain symptoms," says Lynda. "But there's nothing that shows that it works for the long term. Neurofeedback is the most effective intervention I've seen. When Ritalin wears off, you're back at square one, but with self-regulation a child can maintain that state. It's not a panacea, and it's not for every problem. But if it's a problem with attention, this will help. It's a do-no-harm intervention." The big difference, of course, is that neurofeedback takes more time—it requires twenty to forty trips to the therapist's office.

There are a host of low-tech things that can be integrated into neurofeedback treatment. In April of 1999, for example, the results of a well-designed study conducted at Harvard on forty-four bipolar patients were released. In addition to their regular medication they were given omega-3 fatty acids—oil derived from fish. Omega-3 fatty acids are vital to proper functioning of cell membranes. According to an article in the *Washington Post* the experimental group did so much better than the control group that after four months the experiment was stopped mid-way and all of the patients were given fish oil. It is the latest of several very positive studies of the effects of specific types of fat on the brain. Some clients who do neurofeedback training for bipolar or other problems of stability, therefore, also take Omega-3, which is obtainable at any health food store. Some therapists do hair analysis to see which minerals their client might be deficient in. Some research shows that a shortage of zinc can contribute to seizures, for example, and so the client would take dietary supplements.

EEG biofeedback also shows a great deal of promise when combined with other kinds of biofeedback. Douglas Quirk used galvanic skin response and neurotherapy with violent offenders to get them to relax. Les Fehmi says the largest component of attention deficit disorder is anxiety, and relaxation training along with alpha synchrony training can cause symptoms to diminish dramatically in some clients. EEG biofeedback also shows a great deal of promise in combination with other types of biofeedback. Bernard S. Brucker was a Ph.D. psychologist at the Jackson Medical Center at the University of Miami School of Medicine. For the last thirty years he has worked with quadriplegics, paraplegics and people in vegetative states using muscle biofeedback. He has treated an estimated 5,000 people in that time, and the effects have been extremely robust. The concept is simple. Traditional physical therapy asks a patient with a right arm paralyzed in a stroke, for example, to lift a finger. Once they can do that they are asked to move the one next to it, and a little more and a little more. If the person cannot move at all there is nothing for the traditional therapist to build on. Instead of using gross motor movements, Brucker uses faint muscle signals. He might hook up the sensor to the patient's arm, for instance. Even if the patient cannot lift a finger, let alone his arm, he may be able to control faint electrical signals to the muscle, even if he cannot feel them. That in turn recruits neurons in the brain to take over for damaged areas. (This procedure echoes John Basmajian's work teaching people to control a single motor neuron.) Eventually new networks in the brain are created around the old, which replace the damaged tissue. It is not a cure, Brucker said, but everyone he works with gains some movement, and many of the results are dramatic. Brucker often saw people regain the use of paralyzed limbs, often in a matter of four or five sessions. I caught up with him at Rob Kall's Winter Brain Meeting, where he was meeting with EEG neurofeedback people to see how the two different approaches might be wed to make it more powerful.

Brucker agreed that the EEG biofeedback technique as it is used in clinics is in its infancy. "We have a crude understanding of an extremely complex phenomenon," he said. The sensors used in EEG biofeedback measure the electrical activity of hundreds of thousands of cells through the thick plates of the skull. But as the sensing technology becomes much more specific, he says, researchers will know which cells are damaged and people can be taught to control carefully delineated neural networks, even single cells. "Learning, all learning, is about finding and using central nervous system cells in meaningful ways," said Brucker. Bowling or playing the piano or whatever, establish networks between different regions of the brain. These networks are not carefully etched; but actually happen a gross way. What happens after a stroke, for example, is that portions of these networks are damaged. With new, as of yet undeveloped sensing technologies, specific networks near damaged areas—and even specific cells—can be activated or educated to take over for damaged areas. In a very precise way. Far more precise, in fact, than the original learning. "With better sensing technology we will know where in the brain the large EEG signatures that govern motor skills are coming from, and we can teach people how to train those areas, how to activate these networks, at the same time they are doing motor tasks. There is no reason, if the EEG is specific enough, that people cannot accomplish that." It was proven by Basmajian back in the forties.

It's a technique that will not only be applied to all manner of pathology, but to enhanced—perhaps greatly enhanced, functioning. What parts of our brain do we not use at all, that might be engaged to change function? Operant conditioning, not drugs, may create the ultimate designer brain.

New diagnostics have already begun to hone the way neurofeedback is applied. Now QEEG research shows that there are at least three or four different subtypes of ADHD, for example, that have origins in different parts of the brain. "That means we won't be treating all ADHD the same," says Sterman. "The training will become much more specific rather than training everyone the same." Lubar, meanwhile, is experimenting with something called LORETA, low resolution electromagnetic tomography, which, with the use of very fast computers, will allow QEEGs to read the brain's deep cortical and subcortical regions, and may allow the client to train those areas directly, rather than indirectly through the cortex, which should make the protocol much more powerful.

Integration of different kinds of technology into biofeedback may also enhance its effectiveness. Some therapists use a tactile element such as a vibrating cushion or stuffed animal in the client's lap, especially with kids being treated for hyperactivity. As the kids make the right kind of physiological changes, the beeps and visual display are enhanced by the vibrations and information comes to the brain from each pathway.

It's also important that clients find the neurofeedback system engaging so they are motivated to do it. The Wave Rider is a sound-oriented system for heart, brain, and muscle biofeedback. Sound may be the most rewarding element of feedback, says Jonathan Purcell, who oversees product development at Mindpeak, a company in Sebastapol, California. "I set it up so it plays a blues riff," he says. "Every time I get the right reading on my muscle feedback, I get those notes. And the higher I get on the reading, the higher the notes go. Or you record and insert inspirational aphorisms and get a little inspiration for feedback." In the patient's voice, no less.

Thought Technology's Flexcomp is the most sophisticated of the integrated brain wave biofeedback equipment on the market, and probably where the field is headed. It is an eight-channel instrument that allows people to mix and match EEG from the brain, EMG, from muscles and EKG from the heart. It allows the user to customize the audio and visual feedback ad infinitum. Feedback can be provided with the traditional bar graphs and wave form displays, or a client can choose a numerical display and watch the number of microvolts increase or decrease. The system also has a sophisticated programmable collection of computer graphics and music and other sounds that the operator can change. As clients get into the appropriate frequency, they are rewarded with an image of a flower that blossoms or a picture of a sailboat that goes from fuzzy to clear; as their minds wander out of the range, the flower closes or the picture of the boat gets fuzzy. At the same time, they are rewarded with the sound of an ocarina or a clarinet that changes in pitch as they do better or worse.

A great deal of innovation has begun to happen in the field in the last few years. Dr. Alan T. Pope has worked as a research psychologist at NASA's Langley Research Center for thirty years, studying, among other things, the way pilots pay attention in the cockpit. His research, coupled with the equipment the Othmers, Ayers, and others have designed, led him to create a different kind of neurofeedback system called an Extended Attention Span Training System. The game is an off-the-shelf video game, a car racing game. The game has a joystick so the person playing the game can steer the car. But the accelerator is more than a little different than most video games. Instead of a button you push with your thumb, the speed of the car is driven by the brain. "The maximum acceleration you can obtain depends on how well you produce the EEG ratio," that will treat ADHD, said Pope. "It's like the force in *Star Wars*. How well you control your EEG can enhance or impair your ability to play." It's also expected to keep people coming back for treatment, for it is more engaging than other neurofeedback systems on the market.

The system was being tested in a controlled experiment at the Eastern Virginia Medical School at the end of 1999. NASA has entered into a memorandum of understanding with a private company called East3, to market the system.

If you could sit inside a human skull, says Herschel Toomim, a longtime biofeedback researcher, you would have enough light to read a newspaper. His instrument, called hemoencephalography (HEG), has a strap with two light-emitting diodes—one red and one infrared. The light shines an inch through the translucent skull and into the brain and then is reflected back to the surface. Sensors pick up the signal and feed it back to the user. The device in effect enhances blood flow to the brain, creating an enriched environment and, Toomim hypothesizes, a more robust frontal cortex.

Another development starting to take place in the field is the use of virtual reality. Virtual reality is a very realistic, computer simulated three-dimensional world that is displayed to a user in a pair of goggles or a helmet with a computer monitor, and it has found a home in the medical and psychological fields in the last few years.

For example, Brenda Wiederhold, a psychologist and the director of the Center for Advanced Multimedia Psychotherapy at the California School for Professional Psychology, uses a VR system to treat several kinds of panic and phobia disorders, including fear of flying. Patients sit in a real airline seat, with a head-mounted display that surrounds them with a realistic airplane interior. Seats vibrate as the sound of engines are heard. As the familiar palm sweat and rapidly beating heart that accompany panic start to increase, the patient is taught how to control the panic with deep slow breaths, which slows the heart rate and sweating and reduces anxiety. It is easier to make these physiological changes when the sensory experience is so immediate and powerful.

Virtual reality is expected to enhance neurofeedback in a couple of ways. The total immersion and totality of the feedback allows the patient to focus completely on his physiology without distraction and the ability to control brain waves and other physiology comes much faster. Second, it is far more engaging and motivating for the client and it is expected to lower drop-out rates.

The near future of the market undoubtedly includes home use with clinical support. Steve Rothman already has a home-use program for people who do not live near his clinic. As the technique grows, more popular users may hook up to their own electrodes and computer, which will feed their EEGs to a mainframe at the offices of EEG Spectrum or Joel Lubar's clinic over the Internet. Without two computers and software to learn to operate, it will be far cheaper and easier. "At some point, this is going to be like a stereo system," says Dennis Campbell. "Every home will have one. Health insurance companies will demand it." We will all hook up to a computer with electrodes, he says, "and it will monitor vital signs and analyze and compare to your baseline on physiological function. Whenever anything gets out of whack, it will tell you you should do this and that and this. It's like having a spotter when you're doing a difficult gymnastics routine on the trampoline. It's noninvasive, it doesn't involve drugs, and it makes connections when the first signs of imbalance show themselves." Technology futurist Richard Schroth takes it one step further and says people may wear tiny computers that will monitor GSR and EEG and EMG and tell them, in effect, when they need to make an adjustment to reduce stress indicators.

There are some units already specifically designed for the home. The Brainmaster was designed for home use by Thomas Collura, a Ph.D. biomedical engineer who, among other things worked for eight years at the Cleveland Clinic designing brain mapping systems for use in the diagnosis of epilepsy. His two-channel Brainmaster, Collura says, is everyman's neurofeedback machine. "It's like the Apple II," Collura said. "It can do everything the big boys can do," at a fraction of the cost. The software is free and available on the Internet, and the system runs on home computers. The $950 price tag is for the neuroamplifier—the small box that reads the user's brainwaves, amplifies them, and translates it for the computer.

A woman with intense neck and shoulder pain, the result of being thrown from a horse, lies inside the coffin-like beige metal tube of a real-time functional magnetic resonance imager. On a pair of virtual reality goggles there is a flame that is essentially a meter for the neural activity of the brain's perception of pain. In a short time—perhaps just a few sessions—the patient learns to make the flame smaller, thereby extinguishing the pain to the point where it doesn't come back. This is a cutting-edge approach to neurofeedback. In just three half-hour sessions, with the help of the powerful imaging machine, she has learned to snuff out eight years of pain. This feedback is the result of a collaboration between a team of researchers in several departments at Stanford University. In 2005, eight patients in a pilot study were found to have reduced their pain by nearly two-thirds. It is more powerful than traditional EEG feedback because instead of reading signals through the skull it allows the client to go more directly to the brain. Researchers say the technology is extremely promising, and that some patients who have had pain for a long time can reduce or completely eliminate it. "The long-term outcomes look very good," said Dr. Gary Glover, a professor of neuroscience and biophysics at Stanford and head of the team. In fact, it is so powerful an intervention that now one of the prime research challenges, he says, is to see whether the effects of FMRI neurofeedback are reversible. A private company, Omneuron, headed by a former researcher, Christopher deCharms, is developing clinical FMRI neurofeedback treatment—though Omneuron doesn't refer to it as neurofeedback.

The company says the approach could be used to treat addiction, depression, stroke, and epilepsy, and even to achieve peak performance.

"It's wonderful stuff," says Joel Lubar. "FMRI produces very rapid learning. It's extremely accurate and powerful. You can visualize down to the fraction of a millimeter. The technology has advanced to the point where you can target individual cells." In the case of cognitive decline, for example, says Lubar, a few cells might be out of whack. "With FMRI you can learn to increase blood flow and metabolism of these tiny pools of cells in decline in the hippocampus, say. It may slow down the deterioration of the cells." The main drawback, he says, is the cost, which could be more than $1,000 per session. Maintenance of FMRI alone can cost up to $1 million a year.

There are some researchers who believe that the brain is not the best way into the central nervous system at all. The most powerful pathway, they say, is through the heart. According to this model, the human heart is more than a mechanical pump that circulates blood. They believe that with a vast network of 40,000 neurons, neurotransmitters, and proteins that send messages between neurons, and an electrical field 60 times stronger than that produced by the brain, the heart exerts a profound intelligence of its own. This complex circuitry enables the heart to act both independently of, and in concert with, the brain; and researchers at the Heartmath Institute and at other places say the heart can learn, remember, and produce emotions. "We observed," they write in their research paper, "that the heart was acting as though it had a mind of its own and was profoundly influencing the way we perceive and respond to the world. In essence, it appeared that the heart was affecting intelligence and awareness." Information sent from the heart to the brain via nerves, hormones, and other pathways, their research shows, has profound effects on higher brain functions, influencing perceptions, thought processes, health, learning abilities, and especially our emotions and our ability to feel compassion and empathy.

The best kind of heartbearts are coherent. As with Les Fehmi's idea of coherent brain waves, the more coherent the heart rate variability, the more powerful and healing the heart is.

Anger, anxiety, and worry stress the heart and are significant factors in heart disease, including sudden cardiac death. Having a broken heart is not just a metaphor; it is grounded in physiological reality as well.

Heartmath believes that heart rate variability is the key indicator of stress. When positive affect and coherence between heart, brain, and nervous system (in effect the mental and emotional states) are enhanced through a system the Heartmath researchers call an emWave PC Stress Relief System (formerly the Freeze Framer), cognitive activity changes into something positive and happier because has been enhanced positive affect. Everything, in other words, flows downhill from the heart. "The heart is the most powerful generator of rhythmic information patterns in the human body," Heartmath says. "When they are in-phase, awareness is expanded. This interaction affects us on a number of levels: Vision, listening abilities, reaction times, mental clarity, feeling states and sensitivities are all influenced by the degree of mental and emotional coherence experienced at any given moment."

Is it possible? Has a forgotten, somewhat eccentric lost tribe of biofeedback devotees been laboring away, beneath our noses, to develop a powerful way to heal humankind of some of its most intractable social, personal, and economic problems? Have we begun to harness the incredible power of the human mind? Is the brain far, far more plastic than science has proven? Can we use our will to change our biology, to get a personality makeover? Is it this easy, this mechanical, to change the human brain, to erase anxiety and fear and stress as if they were words written on a chalkboard? Has brain wave training returned from a near-death state to spark a true revolution of the mind and brain that will change the whole game: our assumption about who we are, what we have to live with, and who we have to be?

I don't know. I know that in my case brain wave training gave me back a part of my life that chronic fatigue syndrome had taken away and traditional medicine could not treat. Others tell similar stories. But the true scope of brain wave training is a big question that remains to be answered. "Research will tell us for whom it works and under what conditions," says Chris Carroll. "I don't think that this is a panacea that can be used for endless applications with consistent positive outcomes."

Many in the business, however, believe neurofeedback will dramatically change the landscape of health care. "Most of the people in prison are there because of addictions," says Sue Othmer. "They can be helped. Many others in prison are there because of impulse control problems; they can be helped. Many are characterized by ADHD, and they can be helped. Trauma and attachment disorder. They can be helped. Most of the serious mental health problems of our society—violence, learning disabilities, addiction—will be significantly impacted by this training." "It allows you to get your hands on the wheel and steer," said one man who was transformed, "instead of just being along for the ride."

Moreover, the imprisoned right brain could well throw off the shackles of the left, according to Siegfried Othmer. "The last one hundred years has been the century of the left hemisphere," he says, referring to its analytical, rational nature. "It's put science in charge and the rest of us in awe of it, and it defined for us the 'pinnacle of intellectual experience.'" And it diminished the rest of human experience [feelings and emotion] by comparison. It did it at the expense of the right brain, which wasn't even at the table. It's what makes us human, and it didn't even have talking rights. Neurofeedback will help bring about the rise of the right hemisphere to effect a rebalancing. There will be a global shift. The force of this shift will be to give the right hemisphere consciousness standing. And better internal communication between our two hemispheres will be matched in the outside world with a rapprochement between the two cultures."

Biofeedback has been down the road of fantastic, miracle claims once before, of course, in the 1970s. The difference this time is that there are many more people practicing it, using much more sophisticated equipment, getting replicable results and predictable changes. At the very least, these results strongly suggest that the academic bias against behavioral medicine in general and biofeedback in particular be dropped. The sullied reputation that biofeedback got in the 1970s needs to be buried once and for all and the technique honestly evaluated. The research is admittedly thin, but the clinical outcome studies are overwhelming, and those, combined with the predictability of the model, ancedotal information, and the reputations of many of those involved, are extremely compelling. There is precious little else to effectively treat the wide spectrum of problems that neurofeedback purports to treat. Biofeedback needs a second look. Research desperately needs to be funded. Not only to establish beyond a doubt that it works for things besides epilepsy and ADD, but so we can figure out how best to apply it—and so insurance companies and schools and doctors will support it, to make sure it gets into the hands of everyone who needs it.

One of the most exciting things of all is that the field is in its infancy. Clinicians and researchers have been laboring away in their own little corners, with few resources and a great deal of imagination and determination, believing what they have found is the best, and rarely venturing out to talk with others in the field. The story of neurofeedback may, at present, be akin to the story of the blind men and the elephant, in which each man thinks he knows what the elephant looks like from feeling one part of it. Each in reality has only a partial claim on the truth. Each of the neurofeedback pioneers has taught the brain, in his or her own way, to play a piece of music. But no one has brought all of the instruments and players and music together. That grand and complicated and beautiful piece of music has yet to be played.

CHAPTER TEN

Weird Stuff

∞

A young woman shuffled into Mary Lee Esty's office in Chevy Chase, Maryland, one day carrying a scrapbook. She took a Polaroid photo of Esty; put it in the scrapbook, which was jammed with other photos, notes, and scraps of papers; and underneath the photo of Esty wrote the therapist's name and why she was important. The woman was the victim of a serious head injury and had virtually lost her short-term memory. Like the lead character in the film *Memento,* she could not remember what happened long enough to function and had to keep a record.

She had been riding in a car with her boyfriend in California, and when he failed to navigate a curve they plunged over an embankment. Her trauma was almost unimaginable. Years later, in addition to her problems with memory, she had precious little energy and could stay awake only in three 2-hour intervals each day—the rest of the time she slept. She had severe dysautonomia—her body could no longer regulate its temperature and even in the middle of a muggy summer in Washington, D.C., she wore a wool sweater and wool socks and wrapped herself in a blanket. Her speech was halting, and she would take long, painful seconds to speak a sentence. She had lost a quarter of her field of vision, as is common in brain injuries.

After her fourth treatment with the Low Energy Neurofeedback System (LENS)—a total of less than 20 seconds of feedback—she came to see Esty, a therapist with news.

"I did my laundry today," she said.

"Great," said Esty.

"You don't understand. I did my laundry *today*. It used to take me five days."

One morning, further along in treatment, she woke up with a sharp pain behind her eyes, as if, she said, someone was poking the area with a pipe cleaner. The next morning the same thing happened. The third day her field of vision normalized and stayed that way. In the course of twenty-five sessions she went on to regain almost all her function, including the normalization of her body temperature, speech, and memory.

This rapid successful treatment of a problem for which there is no other treatment was accomplished by LENS. As neurofeedback, it is in a class by itself—and some people say it is not even neurofeedback. Instead of just feeding the user back his or her own brain wave, as is the case with most systems, it also feeds back a tiny bit of electrical frequency to the brain. The dose is in infinitesimal shards of a microwatt—a dose of electricity so low it's analogous to the ridiculously low doses of homeopathic medicine. (Homeopathy is the dilution of a substance to the point where there may be no molecules of it left, only, allegedly, a vibrational or energetic "signature"—but somehow it is said to engage the immune system and initiate healing.)

Instead of sessions in which the brain works to create feedback that lasts half an hour or more, the actual feedback from the LENS system lasts 1 second or a very few seconds. One second! It seems absurd. When I wrote the first edition of this book, I considered including the story of LENS and Len Ochs, who founded the system.

He's one of the old guard in neurofeedback, has a good reputation, is smart and articulate, and is full of dry and often self-depreciating humor. But every time I saw him and asked him what he was doing, he would shake his head, laugh, and say, "Weird stuff, really weird stuff." Writing about neurofeedback at the time was weird enough—in the end I felt it would only muddy the water to include a system that was unique and that its own developer routinely referred to as weird.

Och's stuff is still, even by neurofeedback standards, weird. But it is also extraordinarily powerful. Within the field top people have been migrating to LENS over the last few years, using it alone or integrating it with conventional neurofeedback or other kinds of interventions. I think it's time for the technique to receive more attention.

For one thing, practitioners of LENS, some of the most capable in the field, have amazing stories to tell, even by the standards of neurofeedback's impressive results, especially in the complex and difficult area of head injuries. Second, a thorough book on LENS has been published: *The Healing Power of Neurofeedback: The Revolutionary LENS Technique for Restoring Optimal Brain Function,* by Stephen Larsen, Ph.D., who uses the approach. There are also published studies: one by Ochs, Schoenberger, and Esty appeared in the *Journal of Head Trauma Rehabilitation,* and recently the *Journal of Neurotherapy* devoted an entire issue to articles on LENS, including a 100-person study (Larsen and Harrington 2006) and a study showing its efficacy with various animals, including horses, dogs, and cats (Larsen and Larsen 2006).

There is also controversy brewing about LENS, however.

At the ISNR conference at the Town and Country hotel in San Diego in September 2007, I caught up with Len Ochs in his hotel room. I asked him to hook me up for a session on the LENS system, which, like most neurofeedback these days, runs on a laptop computer. He pasted a sensor on my scalp, on the front of my head; fiddled around with his computer; and in a matter of less than five minutes pronounced me done. I had felt nothing. As is the case with other brainwave training systems, I hadn't had to look at a screen or make a race car go with my brain waves or do any other tasks.

"That's it. If you see God tell him I said hello," Len said drily.

We chatted for a while, and then I left. Although I didn't glimpse the deity, the next day I did experience a classic "clean windshield effect," an experience in which the sun seems more golden, colors seem richer, and the Mediterranean breezes of southern California seem most sensuous. Whether it was the single session of neurofeedback or the fact that I was in a gorgeous, sun-washed environment is hard to say. But I feel as if LENS and Len played a role.

Ochs, who is sixty-five and lives in Walnut Creek, California, is a founder of the neurofeedback movement. In high school he was a self-confessed science geek who toured a medical school and saw researchers doing intercranial cell stimulation in rats that he realized was probably biofeedback, which was coming on the scene at the time. He saw an article about neurofeedback by George Von Hilsheimer, a pioneer practitioner, in the Whole Earth Catalog, and started doing it on his own, with an early stand-alone EEG device called the Autogen 120.

Ochs fell in love with operant conditioning. In 1975 he became a counselor at Rensselaer Polytechnic Institute in Troy, New York, and used neurotherapy for his students. "I had every kind of biofeedback I could get my hands on," he said. He also used it himself. "Neurofeedback reduced my drowsiness during the day and improved my sleep," Ochs said. "For every ten minutes of neurofeedback I did I dropped an hour of sleep at night. It was impressive." In 1977 he read C. Maxwell Cade's landmark book, *The Awakened Mind: Biofeedback and the Development of High States of Awareness,* and got interested in Cade's biofeedback device, called a Mind Mirror. A technophile, Ochs built his own with a Heathkit minicomputer kit. He also designed computerized EMG and EEG feedback technology.

In 1990, in response to a request and a $1,000 grant from Dr. Harold Russell, of Galveston, Texas, a prototype machine was born. Small experiments had shown improvements in anxiety and other problems from simple stimulation by light and sound. Neurofeedback was also building a small body of literature showing its power to heal the nervous system. What would happen, Russell wondered, if you combined the two forms? Ochs created EEG-driven-stimulation (EDS), a combination of feedback and stimulation. Instead of just feeding back the brain's electrical signal to the user, the system included light-emitting diodes in a pair of goggles that stimulated the brain. Like neurofeedback, it was used to treat pain, anxiety, and other similar issues. Because of the stimulation, which drives the EEG rather than teaching the brain to make the changes, it was more powerful than other approaches. However, the light proved too much for some people, so Ochs covered it with tape in order to dim it. People were still sensitive to the light and Ochs continued to reduce it's intensity until it was completely covered with black electrician's tape—but even when they couldn't see the light, they could sense it. This was a lesson in some people's extreme sensitivity.

In 1997 Ochs bought a J&J I-330c2—the first neurofeedback instrument with a standard computer microprocessor. It was the inadvertent beginning of LENS.

Ochs was treating a client in Chicago who had fibromyalgia, a debilitating condition of fatigue and extreme muscle pain. This client had done quite a few sessions wearing the light-fitted glasses. Because of his earlier lesson in sensitivity, Ochs wondered if he was overstimulating her, and he took the glasses off and moved them across the room to diminish the effects of the stimulating light.

What came next was a totally unexpected "aha" moment, which defied all of Ochs's expectations. "The wires to the glasses turned out to be an antenna carrying the electromagnetic signal from the computer," Ochs said, still in awe of the unexpected turn of events. When the glasses sat on this client's head the signal was too close, too much, or not the right frequency. Every session had wiped her out and sent her to bed. When he took off the glasses, the wire stretched, and the signal strength got weaker and apparently more appropriate. It was the right amount of frequency. "Instead of her going to bed, which she did after every session, we went for a walk," he said. "It was amazing." The client continued to improve.

Ochs scratched his head. After much pondering he realized that the EEG signal was being processed with a Digital Signal Processor (DSP) chip, an off-the-shelf crystal clock that regulates the timing of instructions to the computer chip. These chips gave off an 11-megahertz signal, and this little bit of frequency was traveling up the EEG electrode wires to the site on the scalp that was being trained. Subsequent examination of the equipment at Lawrence Livermore National Laboratory revealed the therapeutic medium to be a broad band of radio-frequency waves. Together with feedback—and they need to be delivered in tandem, or the approach doesn't work—it was a totally new invention. Ochs called it "a wacko system."

The LENS treatment is based on the dominant brainwave frequency. This could be thought of as where the power is, the frequency where the amplitudes (usually measured in microvolts) are highest.

The LENS approach is known as a "disentrainment" therapy. Instead of amplifying a dangerous frequency tendency or brain weakness that ends in a seizure, it moves away from entrainment. The brain, in essence, is nudged out of its default frequency, or moved, as Tansey might put it, "out if its parking spot." The stimulation is just off the dominant frequency, and it seems to break up existing neuronal patterns and reset them at a different level. Ochs thinks the subtlety and timing of the signal catch the brain and the ego off guard and a person's defenses never have a chance.

It was an accidental discovery, a kind of serendipity similar to Sterman's discovery of the SMR rhythm to immunize cats exposed to seizure-producing rocket fuel (monomethylhydrazine). Chance favors the prepared mind, Louis Pasteur said, and Ochs's mind was prepared to recognize and capitalize on such an anomaly.

Ochs started using the new approach, and its efficacy, he says, was off the charts. "It did it so fast, I was worried," he recalled. "I was frightened. The results were so good." Patients with head injuries, years after being injured, well beyond the time when they would get any natural gains, for example, were getting an 80 percent improvement in the first week of brain wave training—after just seconds per treatment of LENS.

Ochs was tickled and disturbed at the same time. "I always swore never to do anything that didn't make sense," he said. "And this didn't make sense—but something good was happening. It was also eerie and uncomfortable."

I asked Ochs if he has any idea what the radio waves were doing to the brain. "None," he said. "They may effect glial cells or vascular cells, but we don't know." Could it be dangerous? No, he said. The microwattage of the signal is tiny, "smaller than the radio waves around us."

Dr. Stephen Larsen, himself a practitioner, who wrote the first book about LENS, hypothesizes that it works faster than other kinds of neurofeedback in part because the radio waves go deep into the brain. "They penetrate everything," he said. "They go through the whole brain, six layers of cortex, the thalamic and limbic understories, the brain stem. It [LENS] breaks up pockets of coherence, so the brain is not locked into a monolithic response. After a number of LENS treatments the brain becomes more nimble, and it moves quickly though its capabilities."

The big difference between LENS and other systems is "the speed at which it often operates," Ochs says. "Some improvements are like time-lapse photography, where the pumpkin vines grow in fifteen minutes and you go, 'Holy smoke, how did that happen?' Some of the changes are slow and gradual, but some are nonlinear and rapid."

In a session a therapist creates a topographic brain map measuring nineteen or twenty-one sites. ranking the sites according to amplitude. Starting at the left with the healthiest sites—those lowest in amplitude—the treatment progresses to those that are highest in amplitude.

The client is not asked to do anything, merely close his eyes and sit quietly. ("It's a little harder on infants, animals, or autistic kids," says Larsen, "but usually a method is found.") Adults are asked to "be inwardly attentive, and while the session is happening tell me what you experienced—if anything—no pressure being put on the client to have any particular experience." It is suggested to people that if they feel any discomfort or unfamiliar sensations during the treatment they should let the therapist know, and the session will be stopped.

Ochs says that using LENS is not just a matter of hooking someone up. Because of the subtle but sometimes powerful effects, therapeutic skill is required. "There's a careful match of the feedback and the person being treated," he said. "How sensitive are they? How strong?" Changes happen so quickly that the person needs to be told how to deal with them. "What if their life is suddenly not ruled by anxiety when it was ruled by anxiety all of their lives? That's a big change and people need to deal with it."

The LENS approach is different from some other kinds of neurofeedback, in that the process is not training the brain to a specific frequency. "I don't want to tell the brain what to do," Ochs said. "I want to disrupt the malfunctioning of the brain and let the brain reorganize the way it thinks it should. Its like taking the log off the railroad track," says Ochs, "and allowing the train to travel again." That is, he is not telling the train how fast to go; he is simply removing what he calls "the auto-protective mechanisms of the brain" to let it operate on its own. Stress, whether emotional or physical, seems to freeze certain parts of the brain into dysfunction or stalemate, and it persists until gently disrupted.

It's quite an amazing process, and LENS may be at its best with severe closed head injuries. "It treats problems that come on suddenly especially well," says Ochs, such as a head injury or PTSD.

The intervention is being used to treat autism, fibromyalgia, depression, insomnia, fatigue, dizziness, anxiety, Asperger's syndrome, chronic pain, headache, traumatic brain injury, ADHD, stroke, PTSD, and other problems. Because it requires no engagement from the patient, the technique is even being used on animals. Stephen and Robin Larsen use it on horses, dogs, and other animals at their Stone Mountain Clinic in upstate New York. "We even treated a demented rooster named Gustav Klimt, from Brooklyn," said Larsen. "He was incredibly aggressive, attacked man and beast, no matter how big. I only got in one session, but he didn't attack anyone that week."

After their old Aussie sheepdog died, Stephen and Robin Larsen found a dog that had been orphaned by hurricane Katrina. "Aussie Rescue came up with an animal who had been confined to a crate and was lying in its own feces. The dog displayed fear biting and fear barking," said Larsen. "We wanted to see if we could help such a damaged and neglected animal—not only with TLC, but with neurofeedback." They named him Gandalf and did a brain map. "It reminded me of orphans from Russia with reactive attachment disorder I had mapped," Larsen said. "Poor Gandalf was a badly disordered."

Fourteen neurofeedback treatments later Gandalf, said Larsen, normalized. "In the beginning he was noisy, incontinent, untrustworthy, and occasionally aggressive. If you meet him now, he looks and acts—well, like your average sheepdog," said Larsen. "The brain map reflects the changes, and where you had what almost looked like seizures, his brain looks nice and quiet."

There is of course the often-heard criticism that this is the placebo effect. The response of animals would rule that out. The changes also last, practitioners say. And there are responses illustrating that placebo has nothing to do with it.

Larsen tells a story about a Vietnam veteran who was doing a session with LENS. "The patient's eyes were closed," Larsen said. "Embedded in a background recording is one second of stimulation. At the moment of the stimulation (which the vet had no way of knowing) he gasped and struggled in his chair. When I asked what he had experienced, he gritted his teeth: 'Mortar attack, screaming, death.' It was a flashback from his Vietnam war experience."

Using LENS does cause some mild abreaction, something like the Peniston flu, though the problems are transient. Esty said she usually sees the reaction after the first treatment. "Headaches, fatigue, or nausea and dizziness," she says. "They are mild and they usually pass the next day."

When it emerged, LENS threw the field of neurofeedback into controversy—a controversy that lingers—because, it is, as Ochs states plainly and often, a "wacko system." Critics of LENS respect Ochs and his work and others who use the system, but they harbor serious doubts about LENS on a number of levels. "It could be placebo," says Joel Lubar, who admits that the approach confuses him. "We don't know whether (the electrical signal) penetrates the scalp, the skin, the muscle, and other coverings."

It is not, says Lubar, neurofeedback. "It's passive," he says. "The patient has no awareness. Neurofeedback, on the other hand, is an active learning process and the client needs to engage with it." Larsen counters by asking whether neurofeedback need be conscious or even voluntary.

"I don't know what to make of it because I can't explain it," says Barry Sterman. "I can't promote it until I understand it. It can't be magic. You need a theoretical model to show how it might be working. That isn't there yet." On the other hand, Sterman says that he respects Och's work and that Ochs is a member of a small group of researchers Sterman formed, the Society for the Advancement of Brain Analysis.

Both Sterman and Lubar worry that LENS might hurt the efforts of neurofeedback to earn scientific credibility.

No matter how small the signal, Ochs's system is different from standard neurofeedback in a critical way. Neurofeedback has always been able to get around claims that it is a medical device because biofeedback, which only mirrors biological activity, has an exemption. (I believe, though, that if the regulators knew how powerful regular neurofeedback is, they might revoke the exemption.) But the notion of inputting wattage, no matter how small, is different.

Esty's world was upset in 2006 when a client filed a complaint with the Maryland Board of Physicians. Esty had used LENS on a patient who seemed pleased with the outcome. The patient later suffered an anxiety attack, blamed it on Esty, and complained to the Maryland Board of Physicians that Esty was practicing medicine without a license. Her own Board of Clinical Social Workers, however, declared that she had done nothing wrong, and, while a lawsuit drags on, she continues to practice.

"In eighteen years I've never seen anybody damaged," said Ochs. "I've seen people surprised and upset and get restless or sleepy, but these are the stages people have to go through to get better."

Ochs, Esty, Larsen, and the others working with LENS mirror the frustration of the field of neurofeedback as a whole that so powerful an intervention is still not taken seriously by enough people, either by clinicians, researchers, or potential clients. There does seem to be not just a scientific barrier, but an emotional one, that keeps people not only from moving outside their comfort zone but even from thinking about it. Ochs tells a story of a race car driver who came to see him. He sat in Ochs's office and described head injuries he had received in a crash. When he finished, Ochs told him six sessions would go a long way toward helping him. Shocked, the driver stood up and said, "You have just made a mockery out of my injury and everything I have been through." He walked out.

Ochs has just developed a new system, based on LENS, that he is testing. The tests, he says, are extremely promising. He won't offer many details, but he says, "It's not any stronger. But it's exponentially more efficient and precise. And it's as different from LENS as LENS is from traditional neurofeedback."

CHAPTER ELEVEN

A Decade of Change

∞

Walk into Joel Lubar's Southeastern Biofeedback Institute, recently moved to Pompano Beach, Florida, and you are walking into the most mainstream of futures for neurofeedback. Whether your problem is ADD, ADHD, anxiety, depression, or any of a range of other possibilities, you will first be asked to provide a detailed 90-minute social and medical history. Then a therapist will place on your head a tight-fitting yellow, red, or blue cap that looks like a bathing cap. (The color depends on head size.) Conducting paste will be squirted into each of nineteen holes and a sensor and wires will be affixed to each hole. The brain will be measured with your eyes open, with your eyes closed, and perhaps while you are reading or doing another task. The electrical reading from each of the sites creates an EEG fingerprint that is unique to you.

Once the EEG is recorded, the data are shipped to analysts, often including a medical doctor, who transform them into a QEEG and compares this against what's called a "normative database"—an average from a collection of 625 other "Q's" collected from people ranging in age from one year to eighty. It's turned into a color-coded brain map. The diagnosis and the brain map will give those sites on the brain that fall under or over the range of this normative database and guide the neurotherapy to suppress or enhance the brain's signal. The approach can be stated as, "If you see a hole fill it in; if you see a bump, pave it over." (Again, as neuroscience celebrates the idea of plasticity, it's astounding that neurofeedback long ago harnessed plasticity, apparently able to teach the brain to raise or lower the frequency of a single site or several sites and keep it raised and lowered.)

One large school of neurofeedback has moved on from training amplitude or frequency at one site to training coherence—that is, the relationship between sites in the brain. For autism, for example, the area of interest connects the frontal lobe and central motor strip. This section of the brain contains mirror neurons that allow people to mimic emotions for the development of empathy. In autism, it's believed to be underperforming and thus keeping two regions of the brain from "talking" to each other. Brain wave training can help bring the strip online. (Neurotherapy has shown great promise with autism in several solid studies—in fact, it is the subject of some of the best research. In one pre- and post-study, symptoms were reduced by 40 percent.)

Coherence training, advocates such as Lubar say, is a recognition of greater complexity in the brain. "Coherence is a measure of whether brains are communicating with each other or not communicating or overcommunicating," says Lubar. Stress, genetics, and brain trauma are the likely culprits causing hyper-coherence or hypo-coherence.

After twenty sessions Lubar takes another EEG to measure progress. At Lubar's clinic there is also talk therapy, though that is not common among practitioners. "We don't just throw them in front of a machine," Lubar says. "We deal with family dynamics. Sometime the turmoil in the family is so traumatic you don't get behavioral changes." In other words, the effects of new stress outpace the healing of neurotherapy.

A leading Q analyst is Jay Gunkelman, a founder of Q-Pro Worldwide, a company that processes QEEG data and operates a chain of twelve neurofeedback clinics. Gunkelman has been reading quantitative EEGs since the mid-1990s, and he firmly believes that Q's are the future of the field and absolutely vital for appropriate and safe treatment. "Your grandmother taught you not to dive into the water unless you know what's there," he said. "A lot of people dive (treat patients without a Q) and have problems. With a Q you can spot those problems—whether you chose to dive or not is your call." In other words, you can see an anomalous EEG, know there is a problem, and change or otherwise modify the treatment protocol.

Gunkelman recounted a story in which he wrongly treated an autistic child. "I blew the kid up in one session," he said. "He acted out in school. It took us ten sessions to get him back where it started. These kinds of train wreck stories in neurotherapy are not told, but go to the back of the file drawer." There are, for example, he said, three types of obsessive-compulsive disorder (OCD) for neurotherapists: an alpha OCD, a delta-theta mixed version, and a beta version. "There are three totally different neurofeedback approaches for each one," said Gunkelman. If you don't treat each the right way, he said, things can get worse.

Precise placement on the scalp as determined by the Q, says Lubar, is also important. "Moving the electrode 10 percent can make a big difference on clinical outcome," he says. "Just slapping on an electrode doesn't always work. We have 100 billion neurons and a quadrillion connections, trillions of glial cells. It's complex. If you train in a simplified way you might not get good results."

Whereas Lubar treats two sites, a new clinical approach is something called live Z score training. It takes much of the analysis out of training for clinicians because the program figures out the problem and the treatment automatically. It was developed by Dr. Robert Thatcher (who also created a Q database) and Tom Collura of Brainmaster, the Ohio company that makes home training units. It is a four-channel approach that reads the electrical activity of the brain at four sites simultaneously and then works to bring the sites to a normal range, both individually and between sites. It's a very complicated assessment, and training is reduced to a very simple approach—and apparently the brain has no problem training at all four sites at the same time. All the users know is that they are watching a movie—and they need to change focus to keep the movie running. The brain figures the rest of it out.

Although neurofeedback is still a diverse field, coherence training based on a Q is an orthodoxy in ascendance. The new dominant body behind this model is the International Society for Neurofeedback and Research (ISNR), which replaced the Society for Neuronal Regulation. With the FutureHealth meeting in Palm Springs defunct, and AAPB a smaller meeting, ISNR is the premier organization for brain wave training and where the action is for those with a clinical neurofeedback practice. [Barry Sterman has also started a much smaller organization, the Society for Advancement of Brain Analysis (SABA).]

At the ISNR meeting in 2007 in San Diego, I was impressed with the advances that have come to the field ten years after my first visit to a gathering of neurofeedback practitioners. "This is the largest meeting we have ever had," said a beaming Joel Lubar. His wife, Judith, is president of the organization at this writing. More than 500 people had come from around the world to attend workshops and panels on brain wave training. Students in universities have begun to study brain wave training and were present to offer their projects.

The conference was proof that neurofeedback these days is forming into a much more rigorous science, earning its credibility by standardizing diagnosis and treatment and leaving behind its earlier reputation as a field run by "California fruitcakes." There is a strong feeling that neurofeedback, after suffering years of derision and sneering, is about to take off.

This new orthodoxy is becoming decidedly more complex. It can take years of training and study to fully understand a Q. And there are holes in the approach. A client can come in feeling anxious or with ADD, and the problem may not manifest itself in the Q. "If they don't have anything in the Q," says Sterman, who treats a few patients by referral, "I don't treat them, even if they have symptoms. If you can't see the pathology in the Q you can't treat them."

That's one reason why the Othmers say a Q-based approach should not be the new orthodoxy. To Q or not to Q is one of the major fault lines in how professionals think neurotherapy should be delivered. "There are many futures for neurofeedback," says Siegfried Othmer. "Not just one." One of the futures, he believes, is that of Sue Othmer, who has more clinical experience than almost anyone else. She says that training brainwaves to be more coherent on the basis of a brain map simply will not work.

Patient reports are more important, she says: how the person feels at the moment. People are still suffering, she says, even if their condition isn't manifest in the Q. The brain is exquisitely sensitive to frequency, and a patient is a much better arbiter of what's going on, and knows how far training should go. "Norms are frighteningly arbitrary," she says. "Why should my brain be the same as yours when it's functioning optimally? We all function differently. Let's just teach your brain to do what it does better."

Othmer no longer uses the SMR or beta models that are described in earlier chapters of this book. Instead, a big part of her approach is to measure the frequencies at two sites, usually one on each temporal lobe, and subtract the difference. Sue then trains to increase the difference between the two sites, to desynchronize their operation "We train away from synchrony, because when synchrony is beyond what is needed, it's dysfunctional. This approach sheds excessive synchrony. Making the difference between the two sites greater is what we want," she said. "That's what we reward." It's a completely new model, exclusive to the EEG Institute.

"Just doing a general hemispheric training like this takes care of a raft of symptoms that people chase all over the scalp for with coherence training," says Siegfried. Using the Q and increasing the coherence are simply not necessary, they say, except in a few cases. That approach is used to treat everything: migraines, seizures, bipolar disorder, panic, anxiety, and depression.

In some ways it's like the LENS approach. Training slightly more difference between lobes, she says, cradles the brain and allows it to make its own changes, to reset on its own. The SMR beta, protocol says Sue, took the brain where the clinician thought it should go. "We're bringing the brain to a still point where it can reorganize its activity," says Sue. "When it reorganizes on its own, it functions better." That's the problem with the Q coherence training, she says. "It tells the brain what to do."

This "awake state" training the Othmers do, they say, makes the brain more stable and increases its ability to regulate itself. It's about physiology.

After twenty sessions the Othmers' brand of training then moves on, if necessary, to learned behaviors, the emotional stress, the buffeting that comes from growing up with trauma. This is alpha theta training, quieting the brain to a specific frequency and giving it a chance to dwell there, which allows painful memories to bubble up.

Of course what neurofeedback—all neurofeedback—is doing to the brain is theoretical. No one really knows what is happening. People have a sense of it and a theory, but that's where knowledge ends.

Sue Othmer's clinical approach has always rubbed some people the wrong way. Those who follow her clinical model hold her in high esteem as a goddess of the electrodes, but she worries about others who say the EEG Institute is too much of a mom-and-pop operation. Sue has no advanced degree; her technique is not standardized, and it is complicated and difficult to export, critics say. Even people who support the Othmers wish they would do such basics as collecting data on their clients and conducting follow-ups.

The debate over the standards of care in the field is one of several things that led to the next twist in the Othmers' narrative.

The first edition of this book ended the Othmers' story with EEG Spectrum, their first company, declaring bankruptcy. After the bankruptcy the Othmers called ten neurofeedback clinicians who were their friends and were affiliates of the company, and who each put in $25,000 for a total of $250,000. These people believed in the company, and some mortgaged their homes to come up with the cash. They were concerned that they would be outbid, but the company sold at auction for $80,000. The rest of the $250,000 went to "grow" the company. The Othmers took a deep sigh of relief—they had been rescued.

For a while. "Over time a fear of what Sue was doing developed," says Kurt, who went to work for EEG Spectrum in 2000. "The relationship between Sue and Siegfried and the investors went downhill over two years."

However, John and Sebern Fisher, who were part of the rescue, blame most of the split on something they call "founders' syndrome," a common problem when the founders of a business sell it—they simply cannot give up control. "It was not to be that they were going to work in a collaborative way," said Sebern.

And there was concern about the approach. "They tried the free-spirit route," said Sebern. "There are limits to that in a business plan. It's not that it isn't right or good, but we need good science and good practices."

The Othmers, for their part, felt betrayed. For a while Sue and Siegfried become contractors to Spectrum, essentially running their own clinic. But they were not getting to run the company their way, and would not be guaranteed an ownership stake. Once again friendships were dashed on the rocks of the business of neurofeedback. The Othmers worked for Spectrum, but under contract. Tension was high.

By September Kurt told his parents, "I know how to run a company; I know how to market; let me give it a try."

Kurt created what became EEG Info, Inc., and his parents soon followed. Sue now runs the EEG Institute clinic in Woodland Hills and teaches training courses. Kurt is CEO, Siegfried the chief scientist. There are sixteen employees, and in 2007 the company grossed $1.3 million.

In 2003 Siegfried gave a talk at a combined meeting of the Biofeedback Foundation of Europe Society for Neuronal Regulation, near Zurich. A Swiss aerospace engineer, Bernhard Wandernoth, a genius in laser satellite communications, heard Siegfried speak, came up after the talk, and introduced himself. He was visibly excited. "If ten percent of what you said is true, this is a revolution," Siegfried recalls. "He said this is disruptive technology [disruptive technology is a new approach that suddenly overturns the old way of doing things, in this case treating mental and emotional disorders] and I wanted to be involved. He invited himself to the party and became indispensable," said Siegfried. And like others who get involved, Wandernoth saw the power that neurofeedback worked on his own family.

After several iterations the Othmers, in collaboration with Wandernoth, came up with a completely new neurofeedback system called Cygnet that was released in 2007. They met a game designer, Ryan DeLuz, and he also worked with the new system, creating new high-resolution games. The simple, almost prehistoric graphics of the exploding volcano game have given way to sophisticated high-resolution computer graphics.

I tried out the new system on a Dell laptop. At this writing it is the new kid on the block, spiffy and sleek, a giant step from the older and far more cumbersome system. The neuroamplifier is slightly bigger than a television remote; the game graphics are sophisticated; the system is easy to use and easy to reprogram. Its remarkable how far things have come since the first computerized training on the Othmers' kitchen table and the training that I did just a few years ago.

The Othmers are planning to franchise their approach, feeling that the self-regulation revolution is under way. They also run a nonprofit, the Brian Othmer Foundation, to banish all business considerations from one part of the field and deliver services to populations that can't afford them. They continue to work with addictions; and their Cri-Help study, showing the efficacy of neurofeedback for addictions, was published by the *American Journal for Drug and Alcohol Abuse* in 2005. One of their projects is EEG4Veterans, to treat combat veterans with post-traumatic stress disorder.

And they are free from having to worry about the finances of their company, because it is in the hands of someone they trust. "My parents don't want to run a business; they want to do this work and not have to worry about money," says Kurt. "They realized the only way to find the right manager was to grow their own. Me."

Meanwhile, EEG Spectrum still thrives, in Canoga Park. It is run by CEO John Hollister, who came to the company from a position as global marketing director for Amgen, where he helped develop two drugs: one a pediatric vaccine, the other for cancer; one worth $4 billion, the other worth $1 billion. (He was not with Spectrum when it split with the Othmers.) Now he wants to bring that commercial Midas touch to neurofeedback. "I came in very skeptically," he said. "It was a field making wild claims. But I am convinced the modality has so much potential and it's in the hands of so few." Spectrum's approach, he said, is moving in a different, though familiar direction—back to the future. "A lot of people are going back to where it all started," he said, "going back to the work Sterman did." In other words, SMR. "There are other placements that have impact and we're using those," he said, "but clearly SMR has a lot of contact with symptoms, behaviors, focus, and attention issues. It's the core." And it's where the most of research is. The company has about 1,000 affiliates, he says.

He believes a drug-style double-blind controlled randomized study is the key to the future of the field and is raising $1 million for a 240-subject study slated to begin in 2008 and take about nine months. "Once these kinds of studies are done," he said, "insurance reimbursement follows almost instantaneously."

Meanwhile another large project to collect pre and post data, including a six-month follow-up, is under way at nine neurotherapy centers around the world. The study, conducted by clinical psychologist Roger deBeus, is large, with 500 subjects in the Netherlands, Australia, the United States, and several other countries.

Margaret Ayers continued to operate her clinic and made headlines in the *National Enquirer* when she treated the hospitalized comedian Rodney Dangerfield with her "coma box." Dangerfield had gone into the hospital for a heart valve replacement, and surgeons tried a new technique. It sent Dangerfield into a coma, and a month later Ayers was asked by his wife, Joan, to help him. The green light was aimed at Dangerfield's eye, and Ayers began coaxing him. "How does one use humor with the greatest comedian, Rodney Dangerfield?" Ayers said to me. "Well, one of the many things I said was 'Rodney, if I bring you out of this coma, maybe I'll finally get some respect.'" After the first treatment he grinned and eventually opened his eyes. "He was doing well," Ayers says. "He couldn't talk, but his big blue eyes opened and he was very present; he could gesture and he had a big grin."

The next day several friends, also comedians, including Jerry Seinfeld, Roseanne Barr, and Chris Rock, went to see him. The party may have been too much, and he slipped back into his coma. Then a botched enema perforated his bowel and caused sepsis, which ended his life.

In mid-March of 2008 Margaret Ayers died of genetic ischemic bowel disorder. She was sixty-two. Her practice will be continued by her colleague Dr. Penny Montgomery.

The science still lags far behind the business part of the field. Franchising is slowly emerging as a business model. Deborah DuSold is a former commercial real estate broker who was in a bad car wreck. She says that after a couple of years of therapy, she was a mess, with poor memory and poor concentration. Her neurologist told her, "That may be as good as it gets." She found a neurofeedback therapist in Tucson and trained on Zengar, a new, highly regarded system by Valdeane Brown, and the rest was similar to many other stories. "I got everything back—and then some," she said. "Things just kept getting better and better. I had a lot of chatter in my mind. The chatter stopped. There was complete peace of mind. I was happy all the time. I still have emotions, but they go away so quickly it's like water off a duck's back."

She has just opened the first of what she hopes will be a chain of shops called Mindworks Studio in an upscale part of Tucson. Each will be a kind of brain gym, a place where healthy adults go for a central nervous system workout. No diagnoses—people seeking remediation for ADD or other problems will be sent to a clinic elsewhere. "I'm looking for people interested in personal development, peak performance, and spiritual growth," she said. "Therapy's a banned word." "Life coaches," not psychologists or clinicians, will run the shop. She has a range of systems, from Zengar to Jon Cowan's Peak Achievement Trainer to Brain Paint, a new system by Bill Scott that produces astonishingly beautiful fractal images of the brain's activity as it enhances that activity.

There is a trend in the field toward more home use, with systems the user can buy on his or her own. Brain Master is the leading product, built by the longtime biofeedback expert Tom Collura. Many clinicians begin a protocol in their office, and then send the user home with a Brainmaster to finish the training.

There are also several products for home use; the producers include Smart Brain, a company that has adapted NASA's protocol to off-the-shelf Sony Playstation and XBox games for home and office neurotherapy. This approach also allows people to watch their favorite movie as brain wave training—if they wander out of the healthy frequency range, the movie stops. "The biggest issue in neurofeedback is boredom and compliance," says Lindsay Greco, a cofounder of the company. "This technology allows an individual to watch his or her favorite movie or play their favorite game. That creates an immersion and robustness that has been lacking." No first-person shooters, though—it has to be a game with continual motion, which the user has to keep flowing.

In my mind one of the big philosophical questions about the emerging orthodoxy in neurofeedback is the same as the question that arises with antidepressant medications such as Prozac and Zoloft.

Is ADD or ADHD or anxiety or depression really the problem? Or are these conditions symptoms? Mainstream neurotherapy sees the brain as flawed, in need of a tune-up. But there is solid evidence that repressed fear—or emotional stress, as the scientists call it—is at the bottom of many of our most common problems. Because of how the brain processes emotional experiences in early life, everyone has some fear held over from childhood. Some people, of course, have a great deal more. How that fear is expressed is where the genetics comes in. Depending on a person's physiology, it manifests itself as almost anything: anxiety, depression, stuttering, obsessive-compulsive disorder, and so on.

Is getting better enough? Does it matter how we get there? Should we replace the psychotherapeutic model, which emphasizes an understanding of past experiences and self-knowledge, with a mechanical approach that bypasses these things? Some neurotherapy approaches still use talk therapy, but its urgency seems, at the very least, diminished with a mechanical approach. Do we want to understand our early childhood, the role of our parents, and our own physiology? In other words, does content matter? Is the fact that our father or mother withheld love important to know? Do we need to learn specific lessons? Or is wiping away, like words on a chalkboard, the stress that warped our nervous system the only thing that matters? Is an outmoded factory model of education part of the real problem? Will that model ever be addressed if we merely patch 'em up by training the brain to a "normal" range and get them back out on the playing field no matter how dysfunctional the playing field is?

These are some of the serious questions neurotherapy needs to answer as it moves forward.

Personally, I think stress is important to propel people inward, toward some self-knowledge. On the other hand, how much introspection will people undertake? The shortcut provided by neurofeedback doesn't bother Stephen Larsen, a Jungian therapist. "I've cleared out the cobwebs with LENS so I can contemplate the bigger issues," he said. "I don't think years of therapy is indispensable."

Then there is the question of the different places different types of neurofeedback take the mind. Les Fehmi, who has been involved in neurofeedback since its inception, has watched the evolution of brain wave training longer than just about anyone else. "The field used to be people interested in self-transformation," he said. "It's turned into a profession of people interested in remediating symptoms."

"Remediation is too narrow-focused," says Fehmi. "Focusing on the symptom is not enough. Say someone comes in with a fear of driving. When we do synchronous alpha training we're not just treating the fear of driving. We're treating chronic background anxiety. If you take the optimization approach you make gains in anxiety, impatience, irritability, marital relationships, everything. We don't target the symptoms; the problems remediate on their own with optimization."

Adam Crane also thinks beta training is only part of the full spectrum. "After the kid is off Ritalin the elite neurofeedback practitioner will take the kid into alpha theta, to access the creative process," says Adam Crane. "That's the highest order of intelligence. The best intelligence is based on sensitivity, seeing the subtle patterns in yourself and nature. The better the mind, the more it can penetrate that. Low frequency will come to be seen as immensely more valuable than high frequency, especially to functional people."

Alpha is critical for stilling the mind, he said, so thoughts can be observed. "A person actualizes according to how clearly they can observe thinking," he said. "The more you can watch thought, the slower it will move and the clearer it will be. That means clear thinking and it opens the door for more creative thinking. Most people are overthinking so much they are shutting the window to the creative process. Alpha is unbelievably powerful for helping a person learn to open the window and keep it open." A hundred hours of alpha and alpha theta training over the course of a year or two would not only greatly transform the individual, he says, but also transform society.

Elmer Green, at ninety-two the granddaddy of biofeedback, advocates the deepest-state training of all, theta, the 4- to 7-hertz band, the twilight realm. In an unusual three-volume set of books—*The Ozawkie Book of the Dead: Alzheimer's Isn't What You Think*, published in 2001—Green lays out his detailed cartography of spiritual consciousness. Much of it is based on his professed out-of-body travels and his wife's Alzheimer's disease, which he says is a transitioning of the soul into the "bardo," or afterlife realm, as the brain disassembles. "Theta training is for everybody," he says. "Everyone can benefit. It puts you in touch with your higher self. Beta training helps people function, but it doesn't help them get to their high self." His style of theta biofeedback uses a single electrode on the left side of the forehead.

What about, instead of just trying to help people to pay attention better or end their symptoms of menopause or depression, they are taught to go as far as possible? If alpha training takes us into these wonderful transcendent realms, why has the field not focused on offering transcendent experiences, or even spiritual experiences? The hippies and flower children of the 1960s saw a way, with brain wave biofeedback, to change the world—to end war, reduce materialism, improve the lot of the poor, end environmental degradation. I think that promise is still there. What if people glimpsed life beyond the matrix, the day-to-day life or work and play? Would they go back to their quotidian lives, to the mindless consumption, the blur of sex and food and gadgets? Should we settle for "normal"? Or should we settle for just feeling really good? Should we demand transcendence? "It's not enough to take someone from a minus two to a plus two," says Adam Crane. "We need to take them to a plus five."

Whatever happens next, it seems that neurofeedback is here to stay, and "steep learning curve" is the operable term. "The brain is active and wants to learn," says Sue. "The brain sees itself and sees what's doing and wants to learn, and it's riveted." Neurophysiologists have gotten involved in trying to figure out a detailed model for how this might work. "The learning curve in neurofeedback is so steep," says Jay Gunkelman, "that we can't see the top of it."

There is a tribe of people called the Moken, who make a living in the Andaman Sea in southeast Asia. They have the extraordinary ability, after years of diving deep into the sea to gather food, to see clearly underwater. Some scientists wondered whether the ability of the Moken to constrict their pupils to compensate for the depth of the water was the result of some type of genetic anomaly, but other researchers found that they could teach European children to do the same thing.

Although we have broken the code of our genome and explored the farther reaches of our solar system, we have yet to fully understand the level of control we have over our own physiology. The Moken children, the eastern schools of meditation, the studies of yogis by Elmer and Alyce Green in the 1970s, and many other examples show that human beings have a remarkable ability to exert control over what for all intents and purposes seems uncontrollable. The work of John V. Basmajian alone—teaching people to fire small clusters of nerve cells at will, even to play a drumroll—should have ushered in a new paradigm, that we have unimagined control over much of our physiology.

The next great advance in human health will come when we realize we can expand voluntary control of autonomic function beyond the brain, to the heart, the muscles, the eyes, and other parts of our physiology. There is a symphony not only in the brain, but in the body as well.

During the writing of the last edition of this book Sue Othmer said she felt that neurofeedback was like a grand piano on which she had learned "to play a few keys." I asked her that question again, nearly a decade later. We have learned a couple of keys more," she said. "But we still have a long way to go before we know how to play the symphony."

For More Information

∞

- www.isnr.org International Society for Neurofeedback and Research
- www.aapb.org Professional organization for all types of biofeedback
- www.bcia.org Web site for the Biofeedback Certification Institute of America
- www.eeginfo.com Sue, Siegfried, and Kurt Othmer's Company, EEG Info
- www.openfocus.com Les Fehmi and Susan Shor Fehmi's Web site.
- www.neuropathways.com Margaret Ayer and Penny Montgomery's Web site
- www.eegfeedback.org Joel and Judith Lubar's Web site
- www.ochslabs.com Len Ochs's LENS Web site
- www.futurehealth.org Rob Kall's Web site
- www.zengar.com Val Brown's Web site
- www.eegspectrum.com Company run by John Hollister, once owned by the Othmers
- www.brainpaint.com Bill Scott's Brainpaint Web site
- www.smartbraintech.com NASA neurofeedback technology for Xbox and Play Station games

- www.addcentere.com Michael and Lynda Thompson's Web site for their center in Toronto
- www.biocybernaut.com Jim Hardt's alpha training Web site
- www.BFE.org Biofeedback Foundation of Europe
- www.heartmath.com Heartmath Institute
- www.inneract.com Rae Tattenbaum's peak performance Web site
- www.qproworldwide.com Jay Gunkelman's company
- www.imft.org Adam Crane's Mindfitness Web site
- www.thoughttechnology.com Largest manufacturer of biofeedback in the world.
- www.lexicor.com Another longtime neurofeedback manufacturer

Select Bibliography

∞

Abarbanel, Andrew, and James R. Evans, eds., *Quantitative EEG and Neurofeedback* (New York: Academic Press, 1999).

Brown, Barbara B., *New Mind, New Body: Biofeedback: New Directions for the Mind* (New York: Harper and Row, 1974).

Brown, Barbara B., *Stress and the Art of Biofeedback* (New York: Harper and Row, 1977).

Carter, Rita, *Mapping the Mind* (Berkeley: University of California Press, 1998).

Damasio, Antonio R., *Descartes' Error* (New York: Putnam, 1994).

Diamond, Marion, *Magic Trees of the Mind: How to Nurture Your Child's Intelligence, Creativity and Healthy Emotions from Birth Through Adolescence* (New York: Dutton, 1998).

Fehmi, Les, and Jim Robbins, *The Open Focus Brain: Harnessing the Power of Attention to Heal Mind and Body* (Boston, Shambhala Press, 2007).

Finger, Stanley, *The Origins of Neuroscience: A History of Explorations into Brain Function* (New York: Oxford University Press, 1994).

Green, Elmer and Alyce, *Beyond Biofeedback* (Ft. Wayne, Ind.: Knoll Publishing, 1977).

Hooper, Judith, and Dick Teresi, *The Three-Pound Universe: Revolutionary Discoveries About the Brain* (New York: Macmillan, 1986).

Hutchinson, Michael, *Megabrain* (New York: Beech Tree Books, 1986).

Larsen, Stephen, *The Healing Power of Neurofeedback: The Revolutionary LENS Technique for Restoring Optimal Brain Function* (Rochester, Vt., Healing Arts Press, 2006).

LeDoux, Joseph,*The Emotional Brain: The Mysterious Underpinnings of Emotional Life* (New York: Simon & Schuster, 1998).

Lubar, Joel F., and William M. Deering, *Behavioral Approaches to Neurology* (New York: Academic Press, 1981).

Restak, Richard, *Brainscapes: An Introduction to What Neuroscience Has Learned About the Structures and Function and Abilities of the Brain* (Winnipeg: Hyperion, 1995).

Rossi, Ernest Lawrence, *The Psychobiology of Mind-Body Healing: New Concepts of Therapeutic Hypnosis* (New York: Norton, 1988).

Sears, William, and Lynda Thompson, *The A.D.D. Book: New Understandings, New Approaches to Parenting Your Child* (Boston: Little, Brown,1998).

Stein, Donald G., Simon Brailowsky, and Bruno Will, *Brain Repair* (New York: Oxford University Press, 1995).

Stevens, Leonard A., *Explorers of the Brain* (New York: Knopf, 1971).

Index

∞